The Breeding Industry

Its Value to the Country, and its Needs

National Problems

The Breeding Industry

Its Value to the Country, and its Needs

by

Walter Heape, M.A.
Trinity College, Cambridge

Cambridge
at the University Press
1906

CAMBRIDGE
UNIVERSITY PRESS

University Printing House, Cambridge CB2 8BS, United Kingdom

Published in the United States of America by Cambridge University Press, New York

Cambridge University Press is part of the University of Cambridge.

It furthers the University's mission by disseminating knowledge in the pursuit of education, learning and research at the highest international levels of excellence.

www.cambridge.org
Information on this title: www.cambridge.org/9781107423992

© Cambridge University Press 1906

First published 1906
First paperback edition 2014

A catalogue record for this publication is available from the British Library

ISBN 978-1-107-42399-2 Paperback

PREFACE.

IN writing the following pages I have been actuated by the desire to draw closer together two classes of workers in the great field of Biology, the Practical Breeder and the Scientific Biologist.

After some years of work as a student of pure Science, and with some knowledge of the methods and the needs of Breeders, I have formed a firm conviction that the Practical Breeder will gain inestimable advantage from the right application of Science to the industry with which he is concerned. It is my no less strong belief that the field of Scientific Biology will be broadened, the number of workers therein increased, and the means available for their work augmented, as a result of the more intimate relations between Scientific and Practical Biologists which I here earnestly advocate.

The history of all economic industries emphatically demonstrates that their progress and success essentially depend upon the results gained by Scientific research. The Breeding industry is no exception to this rule and our foreign competitors clearly recognise that fact.

Examination of the subject makes it clear that the foundation of any organisation designed for the improvement of this great national industry, must be laid by the Government of the country. Its actual development must, surely, eventually depend upon individual effort; but no amount of individual effort can supply what is needed for a comprehensive understanding of the ever-varying claims and requirements of the industry as a whole, in the absence of which knowledge no substantial benefits can be initiated or maintained.

For this reason I suggest the establishment of a State Department of Animal Industry, organised and controlled by a staff fitted, by their training;—first, to systematically record the condition of the industry throughout the kingdom; secondly, to deal with the many problems Breeders from time to time require solved; and thirdly, to present the result of such work to Breeders in a practical form.

The Breeding industry "is the greatest industry to which Science has never yet been applied," the nation which first does so apply Science will gain great and solid advantage, and it is with the hope of helping my own countrymen to this end that I present these pages to those concerned.

WALTER HEAPE.

CAMBRIDGE.
January, 1906.

CONTENTS.

CHAPTER III. THE NATURE OF THE WORK
REQUIRED FOR THE ADVANCEMENT
OF THE BREEDING INDUSTRY   pp. 74–115

INTRODUCTION.

THE domestication of animals is one of the most primitive attributes of civilisation, and there is no industry with which man has ever concerned himself, of greater importance to his welfare than the Breeding industry.

Look where you will—apart from purely seafaring peoples—the possession of domestic animals and their successful breeding has ever been of vital importance to the success of a nation; whether for the consolidation of peaceful settlement or for the prosecution of war, breeding has always been a matter of primary importance.

Under these circumstances it is a remarkable fact that the history of breeding has never been written, but it is a still more remarkable fact that the application of science to breeding has

H.   I been almost wholly neglected. Perhaps the neglect of the latter is not unconnected with the absence of thè former, but however that may be, we, to-day, concern ourselves but little with either of these things, and we suffer for that neglect to an extent which perhaps few of us recognise.

We have for long been noted for the stock we raise in these Islands. We have for long held the first place throughout the world as breeders. There is reason indeed to believe we do so still, but, and this is the whole point of what follows, our supremacy is seriously threatened.

This industry, which it is not too much to say is one of the most important with which we as a nation are concerned, is about to experience the effects of foreign competition to an extent far beyond anything it has hitherto known.

Other nations are beginning, only beginning it is true, but still beginning to pay serious attention to the science of Breeding.

To us, the result of this new departure, so far, has chiefly been to put money into the pockets of those who have sold, at a high price, much of the best stock they have got. But profits from this source cannot last indefinitely. Some day, perhaps in the near future, these foreign buyers will be in a position to hold their own without any considerable help from our flocks and herds. In one instance at any rate they have already attained that position, and when it becomes general this source of profit will be closed.

But this is not all, nor is it the worst we must anticipate, for, if these foreign Breeders develope the Science of Breeding as they have developed the science of every other industrial pursuit in which they excel ; then, if we do not keep pace with them, they will not only be independent of us, they will beat us with the produce of the stock with which we ourselves have supplied them. This danger is a real one. I believe there are many thoughtful breeders who fully recognise it and I cannot doubt the time has indeed arrived when the future success or failure of the breeding industry in this country must be determined.

The issue is in the hands of the breeder himself, and in urging him most seriously to consider his position in the world now, I would impress upon him the main fact which is to be learnt from a study of the history of the present successful industrial concerns, of every kind, both at home and abroad, namely—the essential part which science plays in the solid development of all practical affairs.

CHAPTER I.

THE BREEDING INDUSTRY.

In the good old days when farming was a profitable pursuit the mainstay of the business was wheat. In those days breeding was largely conducted by farmers whose prosperity was assured, and the gains or losses resulting from breeding were not, as a rule, severally criticised. When from various causes it became profitable to import grain, and the price of home-grown wheat fell, these conditions were altered. It was from wheat the farmer had obtained his profits, and now his energies were directed towards combating, what were to him, the evils resulting from the importation of grain. Breeding of high class stock was comparatively neglected, it became an expensive hobby, it remained chiefly in the hands of men who were not dependent upon the

raising of grain for their riches, and in the hands
of such men it has, to a very large extent,
remained ever since.

The straits in which the farmer as a grower
of grain then found himself, led to the need for
a more scientific conduct of his business; this
led to experiment, and gave rise to popular
movements which have resulted in an official
recognition of his needs. There is a Board of
Agriculture and much useful work being done
by Agricultural Societies, by private persons,
and, recently, by the staffs of the Agricultural
Colleges, Schools, and Institutes founded in
different parts of the country, and by the
Departments of Agriculture established in certain Universities. Now scientific instruction in
the prosecution of Agriculture is provided and is
daily becoming more and more appreciated; now
botanists, chemists, and entomologists are employed to solve the many problems which beset
the agriculturist. He has now comparatively
little difficulty in obtaining expert advice

regarding the quality of his land, regarding the best kinds of grain, seeds, roots, vegetables, or fruit-trees to grow thereon, and the most suitable manures for that purpose. He can readily learn the newest methods for the cure of the diseases from which plants suffer, or the best known methods to be adopted for the eradication of the insect pests which interfere with their satisfactory growth, or with their ability to produce crops of value.

Of course I do not suggest that all these matters are now thoroughly understood, or that the best methods for overcoming the many and great difficulties the agriculturist has to contend with, are already known. The work to be done on these lines is quite illimitable, but, as I have already indicated, the claims of the agriculturist have to a large extent been recognised; the ranks of the scientific workers in this field are rapidly becoming augmented, and, gradually, the great economic benefits which Science can confer upon Agriculture will become more apparent and will be more and more extensively made use of.

The main functions of the Board of Agriculture and Fisheries, in so far as they refer to Agriculture, appear to be; to collect and publish agricultural statistics and intelligence, to promote agricultural education, to administer the sale of Foods and Drugs Acts so far as they relate to agricultural produce, and to prevent the importation or spread of certain diseases among animals. Of its staff, those specially concerned with agriculture appear to be, a Technical Adviser in Economic Zoology, a chief Agricultural Analyst, one or more scientific Assistant Secretaries (whose qualifications are apparently chiefly those required for agricultural problems), a Staff of Inspectors of various grades, and the newly created "Correspondents" all over the country, over 300 have already been appointed, whose duties are specially designed for the benefit of agriculturists.

Besides these gentlemen, certain of the

Agricultural Societies have consulting chemists, botanists, and entomologists on their staff, available for the needs of agriculturists. Special expert advice is also available from bodies such as the West Norfolk Farmers' Manure Association, and indirectly, if not directly, from the staffs of those Universities which include Agriculture in their curriculum, and of the various Agricultural Colleges, Schools, and Institutes throughout the country, to some of which are attached experimental farms.

Surely then the Science of Agriculture is now fairly started in this country, and, notwithstanding the strictly economical lines on which such Science is run and the very scanty supply of money now forthcoming for the purpose, will surely grow.

The breeding industry is, however, in a very different position. Breeding of first-class stock is still largely in the hands of wealthy men who are not necessarily agriculturists, or if they are it is Agriculture which now takes the second place, they are breeders primarily. But the bulk of stock is bred by the farmer, and it is the part he takes which mainly affects breeding as an industry.

Of late years there has been a marked increase in the head of stock in the country.

Perhaps the chief reason for this is the necessary decrease of the wheat acreage. The cause of this decrease is dealt with in numerous papers, journals, and books, and need not be further referred to here; the fact is plain, and the amount of that decrease since 1869, as set forth by *The Times* in an article published August 29th, 1904, amounts to no less than 2,313,073 acres, the difference between 3,688,357 in 1869 and 1,375,284 acres in June, 1904. The land so set free is doubtless used, when it is used at all, for a variety of purposes, but the chief of these would appear to be grazing or the production of fodder crops.

Mr New points out, in a paper read before

the Farmers' Club in May, 1903, that in the last 30 years cattle have increased in this country to the extent of about 1,250,000, and the returns of the year 1904 show a further increase of over 90,000 (*Times*, Sept. 5, 1904). Lord Onslow pointed out in August, 1904, that stock-raising has been the salvation of many agriculturists, and surely there is reason to expect it may come to exercise a still more marked benefit on their fortunes.

Take for instance the increased demand for milk and the stimulus thus given to the dairy interest. This is a comparatively recent development, but it is growing fast. Among the many bad habits that we, as a nation, are continually reminded we indulge in, this growing habit of milk-drinking can hardly be classed, here at any rate is a trade with the further development of which even the most fastidious amongst us must be in accord.

One hears from time to time the fear expressed that the trade is already being over-

done, in my opinion it is in its infancy. But
its satisfactory development will depend upon
the scientific conduct of the business, not only
as regards the treatment of the milk and the
uses to which it can be applied, not only to the
feeding of the cows, but, and essentially, to the
production of more satisfactory dairy cattle.

Take a single instance in confirmation of
my view that the dairy industry is capable of
immense expansion. Graziers tell me there is
a pronounced decline in the demand for fat
meat, that the public won't have it and the
butchers don't buy stock, fat, as it used to be
fattened. This disinclination for fat meat is
associated, if I mistake not, with a greatly
increased demand for butter; since fat is
necessary in some form it is not unreasonable
this should be so, and when our dairy farmers
find the milk trade overdone, there is a vast
business to be done in butter, as our enormous
imports of that form of food clearly show.

Again, if the full significance of the value

of the various constituent parts of milk was generally understood, specially good milk would command a special price from those who could afford to pay for milk of guaranteed quality. The development of such a special trade would benefit both consumer and producer. It is hard to say what class of consumer would fail to take advantage of the opportunity to purchase milk of a high standard of quality, the benefit it confers is so quickly appreciated.

I am told of the substantial success of an experiment on these lines which has been made in a northern city of this country. A certain dairy farmer freezes his milk and supplies it in sealed bottles to his customers, mostly of the poorer class, at a considerably increased price. He finds that the increased value of such milk is recognised to an extent which has enabled him rapidly to increase an originally small business to, I am told, ten times its former size, and still there is a greater demand than he can supply.

A demand for milk of special quality must

result not only in more scientific feeding of dairy cattle but in building up herds of finer quality, and the stimulus thus given to scientific breeding would result in important developments.

The dairyman is not necessarily a breeder. Many in the trade buy cows to milk when they want them, and sell them when they run dry. To such men the business may possibly become less remunerative in the future. Those, however, who breed their own cows on right principles, who build up for themselves a herd of first-class milkers on lines which will admit of still greater improvement, who pay attention to the quality as well as the quantity of milk given, and to the food which is necessary to produce that milk ; such men will reap the benefit of increased trade, as the value of the produce they deal in becomes more generally recognised.

Again, consider the continually increasing amount of meat eaten in this country. Bread has hitherto been considered the main article of food, and everything pertaining to the abundance and price of wheat has occupied the first place in men's minds. It is only comparatively recently that meat has been looked upon by the bulk of men otherwise than as a luxury. In these days, however, meat is a common article of food.

In the Fiscal Blue Book, 2nd Series of Memoranda, etc., on British and Foreign Trade (Cd. 2337, 1904) (*Times*, Dec. 22, 1904), tables are given in which is set forth the amount spent on various articles of food by urban workmen's families. These are classified into families in which the wage earned is under 25*s.*, from 25*s.* to 30*s.*, from 30*s.* to 35*s.*, from 35*s.* to 40*s.*, and from 40*s.* a week. In all these the money spent on meat and bacon is greater than the money spent on bread. When the wage is under 25*s.* bread and flour were put down as costing 3*s.* 0½*d.*, while meat and bacon amount to 3*s.* 2¾*d.*; when the wage is 40*s.* and over bread and flour cost 4*s.* 3¾*d.*, while meat and bacon cost 7*s.* 2¼*d.*; on an average the cost to these working men's families for bread and flour is 3s. 7d., and of meat and bacon 5s. 5d. In these calculations, moreover, no allowance is made for fish.

In a pamphlet issued by the National Hospital for the Relief and Cure of the Paralysed and Epileptic, in a table given of the provisions consumed per year, it is stated that 67,584 lbs. of bread and 78,055 lbs. of meat and fish are eaten. Taking the bread at 1¼d. and the meat (including bacon and fish) at 6d. per lb., the cost of the bread is £352, and of the meat £1951. Many of us buy 4 lbs. of meat to 3 lbs. of bread, and yet, with the 4 lb. loaf at, say 5d., it is still the price of bread we are all exercised about, while the same weight of meat of all sorts costs us 2s. 6d. at least.

It is obvious then that all who are interested in the problem of cheap living, and who is not in these days of wealth? should transfer their attention to the price of meat; it is the butcher's, not the baker's bill which now causes most anxiety to the housewife.

It is odd that this simple fact appears to have escaped the attention it deserves; if it were recognised, one cannot avoid the conclusion, that breeding would in its turn receive attention and a more satisfactory development of the science of breeding follow, as has been the case with agriculture.

These seem to me good reasons for the growth of the breeding industry in the past, and no less sound reasons to expect it will continue to grow.

This essential part of the farmer's work should be more clearly recognised. That such recognition must come there can be no doubt whatever. This country was at one time self-supporting so far as bread was concerned, now it is dependent upon foreign supplies for the staff of life. The English breeder has always enjoyed supremacy over all foreign competitors, and he still holds the first position in the world, but how long will it continue to be so?

We are exporting in considerable numbers

stock of the best blood we have got, our rivals eagerly buy and give long prices for them. This is a matter of supply and demand, it cannot be avoided, those who advocate an embargo on such dealings would at the same time eliminate the stimulus which is at the bottom of our success in breeding. Sir Walter Gilbey laments the transference to Germany of the breeding of a certain class of horse; the stock which enables the Germans to breed this horse was bought in this country, and the business is lost to us. This is a lamentable fact, and it is a clear indication both of the willingness of our rivals to compete against us and of their ability to do so successfully.

But the means to be employed in combating such competition is not repression, as some suggest, it is by increasing the stimulus which governs the success of the industry that our competitors will be mastered, and, as in agri-culture, so in breeding, it is science that must do the work.

Dr Levy, in an article on The Present Position of English Agriculture (*Jahrbuch für Nationalökonomie und Statistik*) argues that England's failure to compete with continental scientific agriculture is due to her disregard of the practical value of science. He asserts that our failure in this field has resulted in our giving more and more attention to the production of stock, in which business he admits we hold a very important position.

Such articles as this one are not needed to assure us that our neighbours are fully aware of the conditions of our business, nor are they needed to impress upon us the keenness with which our failings are recognised and seized upon by our competitors.

If there is one fact more prominent than another in the conduct of all our foreign rivals, it is their determination to make science work for their economic benefit. In all their business science occupies a prominent place; the application of science and scientific method is an essential part of their business organization, and while the results they have thus attained are great, the results they will so attain cannot fail to be much greater. Apply these facts to breeding and I ask again, how long will this country retain its supremacy if it does not also call in science to aid it in the conduct of the breeding industry?

At present the breeder, in our march of progress, is neglected. There are doubtless many reasons for this; the problems are so difficult, breeding is such a subtle matter, the conditions are so varied, the results so contradictory, it seems so hopeless to attempt to grapple with the problems presented, and indeed they are themselves so little understood. All such reasons for neglect may be given, but they are based on fallacy, as I hope to show, and the main reasons why the breeder's difficulties are neglected seem to me to be;—on the one hand a want of appreciation of the importance of the industry, on the other failure to understand the possibilities for improvement and the methods to be employed in order to attain it.

The Breeder then is in a very different position to the Agriculturist. The difficulties he has to contend with are no less important for the success of his business than are those which face the Agriculturist in his struggle for success, the scientific knowledge required is no less. Indeed, the whole of the problems involved in successful breeding are far more intricate than those the Agriculturist has to overcome; they refer to more highly organized living bodies, which are more subtly affected by the very various forces existing in nature and created by the artificial conditions inseparable from domestication. Science has never been applied to these matters, it is not officially recognised as worthy of special study, and the scientific aspect of the many problems which lie at the root of successful breeding are almost wholly neglected.

Let us glance for one moment at the reason

which has brought about the isolation of Science from the Breeding industry. Practical breeders are very apt to look upon the purely scientific aspects of the study of breeding as quite inadequate for their purpose; they not unusually look upon the scientific man as a pure visionary, incapable of assisting them in their practical daily struggle; they are not familiar with the results obtained by scientific research, and the bearing of such upon practical requirements is not plain to them. On the other hand, the experience of past breeders and their own experience enables them to get certain results; they suffer much disappointment, much loss, but this they are accustomed to, and they are disposed to consider both as inevitable.

In the same way, the man of science is apt to overlook the immense value of the store of knowledge which the breeder has at his disposal, he is apt to consider it of very little use to him. He finds difficulty in obtaining a satisfactory account of the breeder's knowledge, he finds his

methods governed by beliefs for which no adequate reason can be given; beliefs, some of which
have their foundation in the experience of many
generations of breeders under conditions which
may be, and some of them doubtless are, entirely
different from those now existing. The scientist
is disposed to consider such beliefs as super-
stitions with which the breeder's reason is biassed,
his experience impregnated, and he is apt to
conclude that much of the evidence to be
obtained from the breeder is untrustworthy.

The attitude of both these men is not without justification, the same difficulty is always
experienced in adapting scientific knowledge and
methods to practical affairs. The task I have set
myself is to show the essential need for overcoming this difficulty, to demonstrate the
immense importance of the subject, and the great
benefits which would follow its satisfactory solution. I am emboldened all the more to undertake
this task, beset with difficulties as it is, because,
as a biologist who has had occasion to consult

breeders continually during the last 15 years, I am impressed with the fact that, in spite of long custom, many practical men are ready to recognise the need for improved methods; they are willing to try those which their financial position enables them to attempt, and are keenly alive to the advantages which careful scientific enquiry promises.

Darwin did much to stimulate the science of breeding, his clear reasoning and the practical value of the great truths he taught has given the modern breeder ample evidence of the important *rôle* science must play in their business. Since Darwin's day much has been done. In the last few years scientific truths have been formulated which will prove of the first importance to practical breeders, and the old attitude of breeders towards science is certainly no longer justifiable.

In his address to the Zoological Section of the British Association in 1904, the President of that Section, Mr Bateson, said :—" Breeding is the greatest industry to which Science has never yet been applied;" and he added, speaking of the present knowledge of the science of heredity, to which he has specially devoted his attention with brilliant success, and of its economic value, " it will be found of extraordinary use." There can be no doubt he is right; none of those who have concerned themselves with the physiology of breeding can doubt that he is right, not only as regards heredity, but in many other branches of the subject of no less practical importance.

So far I have referred only to the science of breeding, I have not yet mentioned the " art " of breeding. I will do so now. In my opinion this " art " is largely responsible for the success breeders have enjoyed in this country. It may be defined as a special attribute of eye and hand which is born with a man, and in its highest development it is a very remarkable gift. I say gift advisedly, it cannot be wholly taught or learned, many have it to some extent, but only those who possess it in perfection have been, and are, the great breeders of the past and the present day. It has been called falsely science, and those who possess it are said to be scientific breeders; but it is *not* science, it is pure art, and of the greatest value; by means of it the infinitely minute variations in animals are recognised and seized upon, and by its exercise successful mating is divined.

But, it may be said, If breeding is an art what need is there for science? The answer is plain, the scientific knowledge of anatomy is no less necessary to the born sculptor than is science to the born breeder. The condition of the breeding industry now is no whit better than was the art of sculpture before anatomy became a science, and by the application of science the breeder will gain no whit less than the sculptor gains by the knowledge of anatomy. As the sculptor of ancient Egypt may be compared with the artist of Greece, so the breeder of the present will, one day, compare with his scientific successor.

CHAPTER II.

THE VALUE OF THE BREEDING INDUSTRY.

I WOULD now make an attempt to give some idea of the magnitude of the Breeding Industry in this country. I venture to think that very few men have any idea of the capital invested in live stock in the kingdom to-day.

The figures it is necessary to go into for this purpose can hardly fail to be tiresome reading; but I am afraid I cannot avoid them if I am to make my story plain, if I am to show that the value of the material with which we are dealing, the stake which the country has in live-stock alone, is represented not by millions but by hundreds of millions sterling.

At the outset of such an enquiry one is met with difficulties which plainly show the impossibility of arriving at anything like a true valuation;

there are no reliable records. But taking such records as do exist, a faint idea may be arrived at of the head of stock in the country, and it seems worth while to indicate, however incompletely, what their value must be.

The only authority to which we can turn for figures is the Board of Agriculture, and there seems to be no doubt that the figures issued by the Board are very incomplete. From evidence I have obtained from various owners of live stock, it would appear that a considerable proportion of them make no returns at all to the Board of Agriculture. A finds that B, C, and D in his district make no return ; he knows this is the case in many districts (the returns are not compulsory), the resulting records are incomplete and untrustworthy and he therefore makes no returns himself.

There is abundant evidence of the truth of this fact, frequent reference is made to it by breeders themselves, and it is stated from time to time by those who have touched upon the subject in writing. But assuming that these figures do represent something like the stock held by agriculturists who are breeders in this country, and under the circumstances the numbers may be safely assumed to be under rather than over estimated; then, adding thereto a very rough estimate of stock used for other than agricultural purposes, a general idea may be obtained of numbers, from which a similarly rough estimate of values, can be arrived at.

Regarding this subject of values a word must be said. It is clearly impossible to arrive at anything like an accurate statement. Various attempts have been made to estimate the total capital invested by farmers, but, as it seems to me, all of them are open to serious objection, for I can find in none of them adequate appreciation of the value of stock.

According to Mr Rew's paper, "Farm Revenue and Capital" (*Journal of the Royal Agricultural Society*, vol. 6, 1895), the total capital invested by farmers, including live stock, land produce, implements, and working capital, was estimated;

(1) By Sir J. Caird in 1878 at £400,000,000 (*Journal of the Royal Agricultural Society*, vol. 14). What basis of calculation was adopted is not known, but it appears probable the amount was arrived at by taking six times the rent.

(2) In the same year Major Craigie (*Journal of the Royal Agricultural Society*, vol. 14) made a calculation based on £8 as the average capital per acre, by which he arrived at £376,000,000 as the total sum thus invested in this country.

(3) Again, in the same year, Dr Giffen ("Recent Accumulations of Capital in the United Kingdom," *Statistical Society*) came to the conclusion that the total amount of farmers' capital was then £667,520,000, and he based his calculations on the income tax returns, at ten years' purchase. In 1890, however, the same author (*Journal of Royal Statistical Society*, vol. 53), taking income tax

returns at eight years' purchase, reduced the total to £521,864,000.

It is with great diffidence I venture to criticise the conclusions these eminent statisticians have arrived at. There is, however, such wide difference in their results, it is clear there must be a fundamental difference in the basis upon which they found their calculations; thus it is perhaps permissible to draw attention to certain points which appear to me to affect the question.

My own investigations induce me to believe the cause of this wide difference is to be found, chiefly, in the different appreciation of the value of stock shown by these authorities.

In the first place, the rent paid per acre is certainly no criterion of the intrinsic value of the stock fed on that land. If therefore this is, as it is assumed to be, the basis of Sir J. Caird's calculations, his results are of no assistance to us.

In the second place, I would suggest that

while records and regular returns may render it possible to arrive at an estimate of the average capital invested per acre, so far as land produce and implements are concerned, the method adopted by Major Craigie gives no sufficient guarantee that the highly variable quality and value of stock held is adequately considered.

Is it to be supposed that the owner of 100 short-horns worth £3000 to £4000, or the owner of a stud of valuable shire horses, estimates the value of his acres in proportion to the value of his stock? May we not with more reason conclude that, from a breeder's point of view, his acre value is represented in accordance with its power to produce fodder of a definite value? which fodder, be it noted, is of the same value, before it is eaten, whether it is fed to prize animals or to mongrels, to thoroughbred cattle or to "cast" ewes.

In other words, Major Craigie's calculation would seem to be based on the assumption that, so far as stock is concerned, their average value is already known, and his justification for this assumption I venture wholly to doubt.

In the third place, Dr Giffen takes the gross income of a particular class, under Schedule B, and multiplies that by an estimated number of years' purchase, in order to arrive at the capital value.

This method would, I presume, give approximately correct results, providing:

A, that the value of stock and the income it returns was known, *or*

B, that the same rate of profit is obtained from stock of various values. In other words, that the annual income derived from a farm has direct relation to the value of the stock kept thereon.

The first of these is, I maintain, not known, even approximately, and it would seem probable this eminent authority has satisfied himself he is justified in assuming the second.

A definite kind of soil is undoubtedly of

H.   3 special value for particular classes of stock. Generally speaking a breeder may be supposed to recognise that fact and to keep the stock which best suits his land and which therefore pays him the best returns. But there is undoubted evidence that such appreciation is not universal, and it follows that there is a wide difference in the percentage of profits gained by breeders on account of this factor alone. My belief is that there is very much more ignorance of the essential importance of this point than is generally supposed. Stock is by no means rarely seen in districts in which it does not thrive well and in which therefore it does not pay well.

I believe that more knowledge of the factors which influence stock, in this connection (it is essentially a physiological question), will result in very appreciable gain to breeders, it is knowledge of very essential importance.

Then again, if Dr Giffen's method is accurate it must be assumed that a cow of the value of

£30 will return double the profit to be derived from a cow worth £15. If not, there ought not to be any demand for the high-priced cow. But is it so? Fashions are not confined to Bond Street. Fashion is evident in the farm also, it penetrates even the meat market, and it has surely to be reckoned with.

For the reasons I have given I suspect that Dr Giffen's results are open to some question, at the same time his conclusions are far more nearly in accord with my own than are the conclusions of any other statistician, so far as I am aware, and I believe this is due to the fact that his method does, roughly, include the real value of stock, whereas the other methods I have referred to do not do so.

(4) The return issued by the Treasury (House of Commons Paper, No. 345, 1885), estimating the 'capital value of farming stock and implements of husbandry' at £300,000,000 in 1885, appears to be based on the above results of Major Craigie and Dr Giffen, but

written down in accordance with the fall in values and the losses in capital employed in agriculture which "it is generally agreed" had taken place since 1878.

This estimate requires no further criticism.

(5) Mr Turnbull in 1893 (Farmers' Club) calculated the gross farm revenue of the country for the year 1892–3 at £180,000,000, and the farmers' capital, 1887—1892, he averaged at £366,743,875.

(6) Mr W. J. Harris in 1894 (*Journal of Royal Statistical Society*, vol. 57) estimated the capital invested by tenant farmers in their business at £352,180,101, and a year's produce at £171,936,927.

In both these two latter estimates the capital value is fixed at just about double the gross revenue.

From my own investigations the conclusion is forced upon me that in all these estimates, with the exception of that made by Dr Giffen, the intrinsic value of stock is practically disregarded.

Finally, two attempts have been made to estimate, separately, the value of farm stock, to which I must refer.

(7) Mr Rew, "Farm Revenue and Capital" (*Journal of the Royal Agricultural Society*, vol. 6, 1895), takes figures for the three years 1891—1893, and averages them as follows:

Cattle and calves ...	11,356,886	at	£8	... £90,855,088
Sheep and lambs ...	32,983,873	„	30s.	... 49,475,809
Horses	2,057,768	„	£20	... 41,155,360
Pigs	3,605,564	„	42s. 6d.	... 7,661,823
Mules and asses ...	700,000	„	£3	... 2,100,000
Goats	1,000,000	„	10s.	... 500,000
Poultry	32,000,000	„	1s. 6d.	... 2,400,000
			Total	£194,148,080

Here we have an opportunity of criticising the value put upon stock. For instance; as a matter of fact considerably over one-third of the number of "cattle and calves" are "cows and heifers in milk and in calf," and only about one-fifth of them are "calves under one year old." Thus an average value of £8 must surely be too low. I will not go into the matter further here, but will treat of it more in detail below.

(8) Mr Turnbull, "Farm Capital and Revenue" (*Trans. Highland and Agricultural Society*, vol. 10, 1898), gives tables of the value of live stock on agricultural holdings in the United Kingdom, estimated according to their estimated average weight, as follows:

Cattle and calves
11,040,700 at 6 cwt. 66,244,200 cwt. at 32s.6d. 107,646,825
Sheep and lambs
31,216,820 at ¾ cwt. 23,412,615   „   40s.   46,825,230
Horses
 2,093,440 „ 8 „ 16,747,520   „   50s.   41,868,800
Pigs
 3,775,630 „ 1 „ 3,775,630   „   32s.   6,041,005
Poultry
40,000,000   3,375,000

Total £205,756,860

So far as this scheme is concerned, while it may or may not be possible to form a fairly accurate estimate of the average weight of animals, the method does not appear to me to possess any particular advantage, and it undoubtedly introduces sources of error which it is not possible to overcome. For instance, to value horses by weight is to totally disregard a great number of qualifications which are the main criterion of their value; in the same way the weight of breeding stock has nothing whatever to do with their value, nor is the value of well-bred stock in comparison with cross-breds to be judged by weight.

Regarding my own investigations, in drawing up the tables given below I have taken into consideration factors which I do not find have been noted elsewhere, but which appear to me to be of essential value.

My endeavour has been to give an idea not only of the value of farm stock, but to give a rough estimate of both the number and value of all stock throughout the kingdom; it is only by such means the vast importance of the breeding industry can be really understood.

I do not claim to have approached accuracy, but it seems to me highly probable that my totals, large as they are, are still considerably below both the number and value of stock in this country.

The agricultural statistics for live stock in the United Kingdom published by the Board of Agriculture in 1903, so far as numbers are concerned, will be found in the table at the end of this chapter, Nos. 1 to 4.

As I have previously stated, these statistics are by no means accurate, but it is quite clear they are under, not over the mark, and I have adopted them as they stand.

The values attached to these items have been estimated in accordance with the experience of various breeders and owners in different parts of the country, who have been kind enough to give me their opinion; I have taken into consideration the figures published by the Board of Agriculture regarding the value of stock exported from and imported into the country, and have borne in mind the large number of valuable prize animals and thorough-bred stock owned in the country, which are doubtless included in these returns.

The remaining items in the table, Nos. 5 to 16, concern animals not included in the agricultural returns. For these, few records are available and my estimates are necessarily very rough, but it has appeared to me essential for my purpose to give some idea of both numbers and values of these classes of stock, and this I have attempted with the aid of various gentlemen who have exceptional knowledge of particular classes, and whose help I would here gratefully acknowledge.

First, with regard to the values assigned to animals included in the agricultural returns.

1. *Pigs.* These prices I feel sure have been fixed too low, sufficient allowance has not been made for animals above the average value. The total works out just under 42*s.* 8*d.* per head. Mr Rew's figure is 42*s.* 6*d.*

2. *Sheep.* The value of the "Ewes kept for Breeding" I have arrived at by comparing the valuation price of flocks taken over by crofters on the West Coast of Scotland, the estimated value of breeding flocks of Lincolns, Suffolks, Southdowns, etc., and I believe 45*s.*

is a low average price for this class. For "others one year and above," which I have fixed at 40s., it is to be recollected that all home rams as well as high-priced rams and ewes bred for export are included in this class.

Those under one year I have taken at 2, 4, 6, 8, and 10 months and averaged the values, making some allowance for pure bred animals, arriving thus at 25s.

The totals show an average value of say 35s. 10½d. per head.

As a check on this estimate I have taken the live sheep imported into and exported from the country with their value, as set forth by the Board of Agriculture in their Agricultural Statistics, 1903.

There are 354,241 imported sheep, valued at £546,063, and 5,579 exported sheep, valued at £67,758. The average value of the whole works out at 34s. 1½d. per head.

But the proportion of exported, that is, well-bred animals, to imported, is 1·57%, and I think there can be no doubt that the proportion of well-bred animals retained in the country is far higher than this. Thus my estimate of 35*s*. 10½*d*. per head cannot be considered too high.

3. *Cattle.* "Cows and heifers in milk and in calf," I have taken as varying between £15 and £23 a head, which are the outside prices given me for dairy cattle. Taking the average at £19, I have added £1 for the large body of thoroughbred cattle in this class which exist in the country.

The "others two years and above," including all bulls, I have averaged at £15; those "one year and under two" at £10; while those "under one year" I have taken as varying between £2. 10*s*. at birth and £7. 10*s*. at 10 months old. The totals show an average value of £13. 12*s*. 3*d*.

When it is remembered that considerably over one-third of the total head of cattle in this country are breeding cows and heifers, and that there are a large number of well-bred animals of this class in the country valued at almost any price from £30 upwards, I think my totals must be considerably below the actual values.

As a check I again take the Agricultural Statistics, 1903. There were 522,546 cattle imported, valued at £9,209,122, and 2,736 cattle exported, valued at £140,244. The average value of the imported cattle works out at £17. 12s. 6d., that of the exported cattle at £51. 5s. 2d.; that of the whole at £17. 16s., that is £4. 4s. a head more than is given in my estimate.

On what principle Mr Rew values all cattle at £8 I am at a loss to understand. Mr Turnbull's figures work out at £9. 15s. a head. This latter figure is closely approximate to the average value which I have set upon all cattle *other than* "cows and heifers in milk and in calf," namely £10, a circumstance which is confirmatory of my view that weight value is no criterion of the value of dairy cattle; the

latter it is clearly impossible to include in either
Mr Rew's or Mr Turnbull's figures.

4. *Horses. Agricultural.* The average value
I put upon these animals is nearly £33. 6s.
Mr Rew and Mr Turnbull both estimate their
average value at £20, a coincidence which
invites the query—how does the latter gentle-
man arrive at the figure of 50s. per cwt., for
horse-flesh?

The price I have put upon "working horses,"
viz. £40, is doubtless too high for purely agricultural work. But from this class are drawn
horses for a variety of work in towns, and the
value of these horses must be considered. For
instance, as I will show below, the London
cab-horse purchased from farmers costs £25.
The omnibus-horse not less than £33. The
lighter cart-horse about £45, while the heavy
dray-horse costs £70 and often considerably
more. Moreover, shire-horses should be included
in this class, animals valuable enough to induce
the owner to insure the unborn offspring for
£100, £200, and even £300.

Again, taking the Agricultural Statistics for 1903, I find 27,266 horses imported, valued at £631,255, average value £23. 3s.; 34,798 horses exported, valued at £734,598, average value £21. 2s. Of these exported horses a large proportion are worn-out animals shipped to Belgium and Holland, presumably for food. Thus of 33,557 horses exported to foreign countries, averaging £19 a head,

A head
19,350 were shipped to Belgium, average £12
9,741 „   „   „ Holland, „   £9
or 29,091 horses, *i.e.* 86·69 % of those exported to foreign countries averaged £11.

On the other hand, the remainder of the total exported horses, viz. 5,707, were valued at £409,223, or £71. 14s. a head.

Of the imported horses, nearly half of them come from Russia, namely, 12,801, while Iceland and Greenland sent 3,073 ; a total of 15,874.

The value of imported horses from each country is not given separately, but, I take it, the above are a very cheap class of horse; those in fact which are imported, as Sir Walter Gilbey points out, because it does not pay to raise them at home; there are no doubt considerably more of this class, but these alone represent 58·22 % of the whole.

Thus it is clear that when 83·6 % of the total exported horses are worn-out animals, and at least 58·22 % of the imported horses are a cheaper class than those raised at home, still the average price of the totals of both of them is above the price adopted by both Rew and Turnbull as the representative price for agricultural horses in this country.

I maintain with some confidence that the figure given by these gentlemen is quite inadequate, and that the average price arrived at by my calculations, viz. £33. 6s. is below, not above the actual value.

Passing now to the other classes of stock, as there are no records available for them, so far as I am aware, I must give a somewhat more detailed account of the methods I have adopted in order to arrive at the figures I have given.

5. *Thoroughbred Horses.* It might be supposed that data for this class of stock should not be difficult to obtain, I have found it exceedingly difficult. I can only give a rough estimate. For numbers the records contained in the *Racing Calendar* and the *General Studbook*, Supplement 1903, have been abstracted.

These figures refer only to thoroughbreds in Great Britain, they are as follows:

Horses entered for engagements 1904.

4-year olds and over .	719
3-year olds . . .	1044
2-year olds . . .	1722
Yearlings	1101
Mares (produce stakes) .	462
Horses not yet included, say	52
	5100

Besides these there are horses not entered for engagements. Of 3163 two-year olds only 1722 are entered; of 3174 yearlings only 1101 are entered; thus there are 4514 of these two classes to be accounted for. Many of them have doubtless been disposed of as failures, but some are still in the stables; say there are 1584

Then there is the breeding stock. The *Stallion Register*, 1904, gives 713 horses at stud, but in the *Racing Calendar* a list of 186 are given which is stated to be incomplete; say there are . . . 200
Mares put to the horse are recorded as . 4456
and foals as 3060
A total for England of . . . 14,400

For Ireland I have no records, but Mr Sargent in his pamphlet on "Sport," 1894, after careful consideration gave figures which indicate that about 25 % may be added for Irish thoroughbreds; say 3600
Making a grand total of . . . 18,000

As for the value of this stock. Mr Sargent gives the value of 7120 English thoroughbreds as £3,560,000, *i.e.* £500 each; and 1770 Irish horses as £265,500, *i.e.* £150 each. A total of 8890 horses worth £3,825,500, say £430 each.

The blood stock sales for 1903 resulted in the realisation of £387,870 for 2177 horses, of which a considerable proportion were failures sold for a few pounds, which works out at say £178 a head.

If we take 7120 of the English horses at £500, and the remainder, 5696, at £170, a total of £4,528,320 is reached; and if we take 1770 of the Irish horses at £150, and the remaining 1830 at £50, £357,000 is obtained. Added together these sums amount to £4,885,320.

At £270 each for the whole 18,000 a total of £4,860,000 is arrived at.

In the table presented at the end of this article, the prices given, with the exception of horses not entered for engagements and the Irish horses, were arrived at before I saw

Mr Sargent's pamphlet. I cannot claim any sufficient justification for the values fixed upon by me, but the total thus reached, viz. £4,879,000, is in such close agreement with the figures obtained by following Mr Sargent's valuation and the prices realised at blood stock sales, that they may perhaps serve for the very rough estimate which alone I can hope to present.

6. *Army Horses.* A committee of the Hunters' Improvement Society (*Times*, Dec. 9, 1904) reports that in time of peace the Government purchase 1,500 to 2,000 horses per annum for the use of the army in this country. If that be so, taking the supply at 1,700, and allowing a working life of say 15 years, we cannot assume that the army possesses more than 25,000. Taking the price given as £40 per head that amounts to £1,000,000. Exact figures are not available from the War Office, but I gather this estimate is substantially correct.

It is perhaps of interest to note that in a lecture given by Major Moore at the Royal

United Service Institution (*Times*, Dec. 8, 1904) it is stated that the supply of horses capable of use for the army in the British empire is 8,000,000.

The statistical abstract for Colonial and other possessions of the United Kingdom, 40th, No. 1903, returns the number of horses in certain of these countries in 1902 as 2,346,158. For certain countries no returns were made for that year, but if the returns last made for these countries be added, viz. 1,505,167, a total is reached of 3,851,325.

Deduct this total from the estimate given above, 8,000,000. Then horses in the United Kingdom fit for military purposes should amount to 4,148,675, which at £40 a head represents a sum of say £165,947,000. The interest attaching to this estimate will be apparent when the total values of horses are examined.

7. *Harness Horses.* In order to arrive at a rough estimate of the harness horses in the kingdom I have taken the Inland Revenue

return of licenses for hackney and other carriages in Great Britain (46th Report, 1903). The total, 584,591, includes machine-driven vehicles, but does not include carriages drawn by horses in Ireland. Taking into consideration the fact that horses are, as a whole, cheaper in Ireland than they are in England, and that the machine-driven vehicles included in the Great Britain returns are largely vehicles which replace omnibuses (these I will show later require a large proportion of horses per vehicle), I have balanced the one against the other and taken the total for Great Britain as representative for the United Kingdom.

The number of four-wheeled vehicles for two or more horses is given as 190,338, those of four or two wheels for one horse as 394,253, a total of 584,591. The return of the Commissioner of Inland Revenue, although it does apparently distinguish between public and private vehicles of four wheels, returns all two-wheeled carriages under a single heading. The four-wheeled hackney carriages (public conveyances) are returned at 133,164, while the other four-wheeled carriages amount to 128,296, very nearly equal numbers.

The total for two-wheeled carriages is 323,131, a large proportion of which are probably traps in use in country districts. If we take therefore 200,000 of these as private conveyances, and the remainder, 123,131, as public conveyances, we shall not, I think, be over-estimating those owned by private individuals. Thus, of the total carriages licensed, I estimate 328,296 are private conveyances and 256,295 are public conveyances (584,591).

In estimating the number of horses used for harness work I have considered the following points. Of the private carriages 57,174 are for two or more horses, and 271,122 are for one horse. Taking into consideration the fact that, while more than two horses are kept for the former class some of those animals are used also for single horse carriages, I have estimated the horses in accordance with the vehicles returned. Thus, 57,174 carriages at 2 horses per carriage and 271,122 carriages at a single horse per carriage, amount to 385,470 horses.

For the 256,295 public conveyances however a more liberal allowance of horses must be made. From Scotland Yard I learn, that for the year 1903, licences were granted to 11,403 cabs and 4,766 omnibuses, a total of 16,169. Allowing two horses for each cab and 12 horses for each omnibus, which is I believe a fair estimate, the total horses for these vehicles amounts to 79,998 or just about 5 horses per vehicle. The proportion of omnibuses to cabs in London is 1 to 2·6, but as omnibuses are doubtless at present in excess in London, and as many of the provincial omnibuses are now replaced by electric tram-cars, I estimate that one-tenth of the whole may be taken for omnibuses (25,629·5) and nine-tenths for cabs and carriages let out on hire (230,665·5).

Taking 12 horses per omnibus a total for

this class of horse of 307,554 is attained, and taking an average of 2 horses per cab or carriage let out on hire, a total of 461,332 is reached, making a grand total of 768,886 for the horses used in public conveyances; an average of 3 horses per vehicle, which is surely below the actual figures.

In order to arrive at a rough estimate of the value of these horses I have taken omnibus horses, which cost in London from £33 to £35, at the lower figure; and cab horses and horses for carriages let out on hire at £25, which is the sum estimated by a large dealer in cab horses as the average price of the horses he supplies. This price may be thought high, and perhaps for cabs alone it is high, but the horses used by jobbers for the carriages they let out on hire, which are included in this item, cost very much more; and the opinion of those in the trade whom I have consulted is, that the price I have taken is certainly below, not above the average.

Carriage horses on the other hand cost from

£30 to £200. In estimating an average price I have taken into consideration the large number of conveyances, included in this class, which are used in country districts and horsed by inferior animals to those seen in the large towns. I have therefore taken £60 as an average price.

The total thus obtained represents the value of the animals in use for harness work: we have still to consider the breeding stock. If a horse works 15 years on an average, we require, in order to keep up the supply, 76,957 horses per annum. Say 10% of the breeding mares of this class of horse fail to breed each year (amongst thoroughbreds the proportion is over 40%) and 5% of the young stock produced are failures or die, then 88,499 mares are required for this purpose; together with say 2,950 stallions, that is a stallion per 30 mares.

Thus, allowing nothing for accidents, it is clear that not less than 91,449 breeding animals are required to keep up the supply of harness horses.

According to the Hackney Horse Society's *Stud Book*, vol. 20, 1903, the stallions entered for the year numbered 306, mares and their produce 749, while the stallions and mares returned as the property of members of the Society in Dec. 1902 numbered, for horses 2,747, for ponies 406.

Canada and other countries are undoubtedly the breeding grounds for a proportion of the horses used in our public conveyances, though I imagine by no means to the same extent now as was the case a few years ago ; and, as Sir Walter Gilbey has pointed out, a large proportion of our best carriage horses are imported from Germany. Thus a considerable proportion of the breeding stock for harness horses is not owned in this country, and, apart from hackneys, it is probable that a perceptible proportion of the stock we do own is included in the returns of the Board of Agriculture, amongst horses used for agricultural purposes.

Under these circumstances I have omitted all

mares, and have taken half the estimated number of the sires required, as the proportion owned in this country to be accounted for, viz., 1,475 head. The price I have put at £100 per head. To this must be added half the foals and young stock which, on the supposition that they are broken as three-year olds, requires a further addition of 115,435 head; and these, on the supposition that it is chiefly the poorer class we breed in this country, I have taken at the low average value of £15.

A special committee of the Hunters' Improvement Society reports that it is said over £2,000,000 per annum is paid to foreign breeders for horses imported into this country, and as it appears probable the majority of these horses are used for carriage work, the conservative nature of the estimates I have given above is thus confirmed.

8. *Light Cart- and Heavy Dray-Horses.* Of this class of horse I calculate that there are considerably more employed than are used in

carriages and public conveyances. When it is remembered what a very large number of such light horses are used by shopkeepers in their business in towns and in the country, and to what a very large extent merchants in all large towns, railway companies, carriers, coal-owners, brewers, and manufacturers all over the country rely on horses for the movement of goods, I think there can be no doubt this conclusion is justified.

While a single horse is sufficient for the light carts used by shopkeepers and some carrier carts, the heavy carts often take two horses, drays almost always two, and sometimes, in the country districts especially, three.

As an example, for heavy horses; a single railway company owns, according to a recent report, about 5,000 horses, the average cost of which is from £70 to £75, while the number of both light and heavy horses employed in the work of that railway reaches, I am informed, the enormous total of 120,000 to 130,000.

From this we may, I think, safely calculate that the 46 railway companies enumerated in the stock and share list I have before me *own* not less than 50,000 horses.

As an example for the lighter class of horse, one large company of carriers owns about 3,000, the average cost of which is over £45, and they buy their animals remarkably cheaply. On the supposition that the 120,000 horses employed in the work of this one railway is an exceptionally large number, we may I think safely put the total horses used by railway companies and carriers at not less than 1,250,000.

This may appear a large number but in reality I believe it is considerably below the actual number thus used. It will be convenient to examine details of the work done by horses in the brewery trade, for information regarding which I am indebted to a friend who has exceptional facilities for obtaining knowledge.

We may say 34,000,000 barrels of beer are brewed per annum in the British Isles, and that

5 barrels weigh a ton. A pair of horses will distribute in London 15 barrels, that is 3 tons, per day; or, taking 300 working days per annum for this pair, 4,500 barrels a year (*i.e.* 900 tons). Thus it would take 7555 pairs, or 15,110 horses, to distribute in London all the beer brewed.

But much is delivered in country districts by single horse carts. The highest record given me for town delivery for one horse (short journeys), is 3700 barrels per annum; that means that 9189 horses would be required to deliver the total output thus. On the other hand, 1400 barrels per annum is the lowest record for country work (long journeys), which means 24,286 horses would be required to thus deliver the total output. The bottle trade of course will require a much larger proportion of horses. But taking these figures; if 10,000,000 barrels were delivered in the country, it would take 7143 single horses to distribute it; and if 24,000,000 barrels were delivered in towns, 6487 single horses would be required for the purpose; a total of 13,630 horses. But a pair of horses will not distribute as much per head as a single horse, and many pairs are used for the purpose, we may therefore say, in round numbers, 15,000 horses are used for the distribution of beer in the British Isles. If this is correct, one horse will be required for say 450 tons per annum.

Now, returning again to the railways. One large railway company moves from 40,000,000 to 50,000,000 tons of merchandise and minerals per annum, say 45,000,000 tons.

If three-fifths of this is moved by horses both to and from the rail, 54,000,000 tons are thus carried, which at the rate of 450 tons per horse per annum will occupy 120,000 heavy dray-horses.

But a considerable amount of the tonnage moved is done by lighter horses which will not draw so much; thus the 120,000 horses employed by this one railway will be readily accounted for if not more than half the total tonnage is moved by horses. At the same rate the 1,250,000 horses which I estimate are engaged in the railway and carrying trade, will be required for the movement of not more than 562,500,000 tons per annum; and it must be remembered that goods are moved frequently twice and often more times before they are ultimately distributed. The final distribution of goods, carried out by smaller horses, in much smaller quantities, over a much wider area, will, I estimate, occupy roughly an equal number of horses.

Taking the total light and heavy cart-horses used for other than agricultural purposes at 2,500,000 I will estimate 40% are heavy dray- and 60% are light cart-horses. The price of the former I will put at £60 and the latter at £30.

The breeding stock of this class of horse is no doubt chiefly owned in this country. Taking fifteen years as the working life of such horses, and allowing 10% for failures, it is necessary to employ 73,332 heavy dray- and 110,000 light cart-horse mares per annum; that is, 183,332 mares are required for the production of 166,666 foals.

Horses used for agricultural purposes total

1,477,991. In order to keep up the supply, allowing a working life of 15 years, 98,533 foals per annum are required ; allowing 10 % for failures a total of 108,386 mares are required. When these are added to the above total of 183,332, a grand total of 291,718 mares is reached.

In the *Shire Horse Stud Book*, vol. 26, 1904, the number of effective shire horses registered is 22,949 ; so that if each sire got, say, 13 foals per annum, the number of foals estimated above would be obtained.

Again, however, I will omit mares, some of which are no doubt included in the agricultural returns while others are working, and will add 14,100 stallions, that is one for each 13 of the 183,332 mares required ; and will price these at £150 each.

Of the young stock, some will be included in the agricultural returns for unbroken horses and foals, totalled at 591,953. I will assume this is a three-years' stock, and represents an annual supply of 197,318 foals. As already shown, 98,533 of these are required for agricultural purposes, leaving 98,785 for the town supply; but as 166,666 are required annually for the up-keep of town horses, there is a balance of 67,881 still to be accounted for, a three-years' stock of which will total 203,643. I value these at an average of £20.

9. *Hunters.* *Horse and Hound* gives a list for 1904–5 of 228 packs of Stag and Fox-hounds in the United Kingdom.

Mr Sargent in his pamphlet on "Sport," 1894, gives 209 of these packs and adds for Harriers 145; which would bring my total up to 373 packs. The same author estimates there are 100 hunting-men with three horses each for each pack, and he values the horses at £80 per head. To these he adds 3,600 servants' horses, which he values at £50 per head. This works out at 109,800 horses of a value of £8,676,000.

I have no claim whatever to criticise Mr Sargent's estimate, but following the plan I have adopted all through, to understate rather than overstate, I will average the number of sports-men to each pack, including servants, at 75, allow 3 horses per man and estimate their value at £75, the total horses therefore will be 83,925.

For breeding stock I will take the average life of the hunter at, say 8 working years, in which case the annual supply required will be 10,490. The production and rearing of hunters is usually attended by more loss than I have allowed for harness horses, in all probability 20% is not too much to allow; but some of the young stock which are not fit for hunters may be used in other ways, so I will again make only a 10% allowance for loss; in which case 11,539 mares will be required for the purpose, and, at 25 mares per stallion, 461 sires. The mares I will price at £40 and the stallions at £250 per head.

The young stock broken, say as four-year-olds, will amount to 41,960, on which I would put an average price of £30.

The total value according to my calculation is thus over half a million less than Mr Sargent's total for horses in the field.

10. *Hacks.* Hacks of various kinds I will

estimate roughly at half the number of hunters, say 40,000, and value them at say £40 per head. With an average of 12 years' working life, 3,333 mares will be required to produce them, adding 10% for failures, say 3,666 mares; these I will value at £25, and say 160 stallions at £150 each. The young stock, broken at 4 years, will amount to 13,332, and I will value them at an average of £20.

11. *Polo Ponies.* Mr Sargent calculated that in 1894 there were 500 players of polo with three ponies each, which he valued at £50 a-piece. I will take the same number, 1500, for playing ponies; calculate that 250 mares valued at £25 and 12 stallions at £100 are required to breed the annual supply, which, if broken at 4 years old, will amount to 1000 young stock of an average value of say £25.

12. *Other Ponies, Asses, and Mules.* These I will put down as 800,000 in number of an average value of £3, which is surely low enough.

13. *Goats.* I will take Mr Rew's figures for these : 1,000,000 in number at 10*s.* a head.

14. *Dogs.* In the Annual Report of Proceedings under the Diseases of Animals Act, Board of Agriculture, 1903, dogs for which licences are granted in Great Britain are returned at 1,575,583

Dogs exempt from licences . 348,460

While in the Report of the Department of Agriculture, Ireland, on the Diseases of Animals Act for 1903, dogs for which licences are granted number 448,750

A total of 2,372,793

The number of stag- and fox-hounds in the United Kingdom, according to *Horse and Hound* for 1904–5, amounts to 16,447 hounds for 228 packs, that is, 36 couple per pack. If the 145 packs of harriers have 20 couple each, that is 5,800, the total hounds will be 22,247, of an average value of say £8.

I estimate there are 3,500 greyhounds in the *Greyhound Stud Book*, which we may take to be of an average value of £15, and perhaps, as Mr Sargent says, there are five times that number, that is, 17,500, not in the *Stud Book*, worth say £5 a head.

The sporting dogs Mr Sargent estimates at about 650,000 of an average value of £2, which I think is certainly low enough.

The dogs exempt from licences may I think be put down at the same relative value, £2, and there are 348,460 of them.

When all these are deducted from the total, 1,331,086 remain. Of these I will take 331,086 as show dogs and others of good quality and breed, and will value them at £5 per head; the remainder of 1,000,000 I will put at 10s. a head.

15. *Cats, ferrets, rabbits, etc.* I will put down as 10,000,000, worth 1s. a head.

16. *Poultry and other Birds.* The number of these will be very large. There are 47¼ millions of farm stock; farmers will probably keep, on an average, as many turkeys, geese, ducks, and fowls as stock; besides which enormous numbers of poultry, pigeons, cage-birds, etc., are kept by others than farmers. I shall put them down as 50,000,000, worth 2s. a head, which is a low price when the value of the larger poultry, table fowl and prize birds are considered.

From the total thus arrived at a deduction must be made for depreciation. From this calculation the young stock are of course excluded.

Roughly the stock concerned may be valued at £350,000,000, a portion only of which will be beyond their prime while another portion will be still undergoing appreciation.

Adult farm stock, exclusive of horses, is valued at say, £110,000,000. The major portion of this class of stock is used for food at the time when it reaches its highest value, only a small proportion of depreciating value is retained.

Allowing therefore an average of 20 % for depreciation on one quarter of these animals, the sum to be deducted amounts to £5,500,000. For the rest of the adult stock, valued at £240,000,000, if one half are beyond their prime and their depreciation be averaged at 20 %, the amount to be deducted therefrom will be £24,000,000.

Thus a total of £29,500,000 deducted for depreciation leaves a net value for stock of £443,476,500.

The complete absence of all reliable data

makes any attempt at accuracy wholly impossible; I have not hoped to attain it, and I don't suppose I have done so within perhaps £20,000,000 to £30,000,000 or even more. All I have attempted is, to give such a broad idea of the number and value of live stock in the kingdom, as the careful consideration of evidence I have been able to obtain, permits. I have taken the utmost care to avoid exaggeration, and in this, at any rate, I have reason to think I have succeeded.

When it is recollected that the Board of Agriculture Returns are below, may be 10% or even more below the correct figures; when it is recollected what a large proportion of the people in the country, farmers, dealers, shopkeepers, farm-labourers, working men of various kinds, and gentlemen's servants, make their living in one way or another by means of stock ; when it is recollected what a very large number of valuable animals there are in this country, as shown by a sale of yearlings at Newmarket, the prices obtained at the dispersal of a herd of

Shorthorns or a flock of Southdowns, the value of a successful horse on the turf, of a good hunter, polo pony, pair of carriage-horses or cart-horses, of a couple of pointers, a spaniel, a bull-dog or lap-dog, etc. ; when such facts are borne in mind I do not think there can be found justification for objection to the final figures I have arrived at on the score of excess ; and yet they show a total sum of nearly £450,000,000 invested in live stock in this country.

When to this is added the capital necessary to provide both buildings to house the stock, land on which to grow their food, barns, machinery, vehicles, harness, and attendance, the total becomes so gigantic that I am surely justified in asserting;—We have here an industry of enormous importance to the country, and one which merits far more attention than has ever yet been accorded to it ; an industry to which, it must be remembered, " Science has never yet been applied."

CHAPTER III.

THE NATURE OF THE WORK REQUIRED FOR THE ADVANCEMENT OF THE BREEDING INDUSTRY.

I HAVE already briefly outlined the position of the breeding industry, and have given some idea of its immense value and importance to the country.

I propose now, very briefly, to indicate the nature of the work which I believe is necessary to place breeding on a sound basis and to assist the industry along a sure road of progress.

That such a task is beset with great difficulties no one who has considered the question can fail to recognise. I am fully conscious of that fact, moreover the difficulty is increased rather than diminished by the necessity to write briefly.

The scientific reasons which form the foundation of the suggestions I shall make can only be glanced at, and the details which weld the whole into a workable scheme cannot be fully enumerated here. A comprehensive treatment of such matters would be out of place in this book. In spite of such abridgement, however, I hope to make my meaning clear, and to demonstrate my belief that the welfare of a great industry, an industry closely bound up with the welfare of the country, is deeply concerned.

Although much valuable work on special branches of the subject is done by private individuals and much more could be done by comprehensive private organisation, it would be futile to attempt even that without adequate and assured endowment.

All breeding work must be slow, the effect of experiments can only be measured at long intervals, results only attained with infinite pains and patience by means of experiments on a sufficiently large scale, carried on for a series of years. Without assured endowment it might at any time be necessary to abandon work of the utmost value when only partially completed (breeding results cannot be manufactured), and years of labour would thus be lost for want of the means to carry it to a conclusion.

On the other hand statistical work must, if it is to be approximately correct, be conducted officially, and all that portion of the work relating to diseases which is closely bound up with the physiology of breeding, undertaken by a Government department which has power to enforce regulations.

After all, the matter is a national concern, and for the sake of uniformity I will conclude, that the Board of Agriculture could be organised to include a Department of Animal Industry.

The work of such a Department would then consist of three main branches,

1. Records,
2. Research,
3. Administration,

1.   *Records.*   It is primarily essential that an accurate knowledge should be gained of the

conditions of the industry throughout the kingdom, and means provided for keeping such records up to date. Without such knowledge neither research nor administration can be conducted to the best advantage; with it the work of both can be performed with the least expenditure of energy and money. If organised on a sound scientific plan such records would be of inestimable value; they would clearly show the weak as well as the strong places, would indicate lines on which investigation and administration would most benefit the industry, and permit of a great reduction in the cost of administrative work. The Records Department would, in point of fact, serve as an Intelligence Department, without which no other branch of any Office of State can be reasonably and efficiently conducted.

As I have already shown, there is no authentic record even of the numbers of stock. In one of the earlier reports of the Royal Commission on horse-breeding, it was urged that such knowledge was essentially required for horses; it is no less necessary for the true understanding of a great variety of facts relating to all classes of stock, and also in order that it should be possible to institute comparisons and form generalizations which are of vital importance to breeders and to the trade.

As an example, take the official figure of the breeding ewes and lambs, quoted in *The Times*, May 9, 1904. Figures are given for "the ewes kept for breeding," and "other sheep," which latter are divided into "one year and over" and "under one year (lambs)" with "totals" for the whole, for each year from 1893 to 1903. These figures relate to the 1st of June, in each year, and the writer draws the deduction, which the official table is apparently designed to enable one to draw, "that the lambs returned in any year are the offspring of the ewes which were returned in the preceding year."

In 1898 the "ewes kept for breeding" are given as 10,137,932, and in 1899 the "lambs"

are given as 10,737,317 (the greatest number of
lambs recorded during the years under review),
which would show 105·91 lambs per 100 ewes.
But, taking the produce of all the ewes recorded
from 1893 to 1902 (100,502,187) as represented
by the lambs recorded from 1894 to 1903
(102,346,808) the proportion works out as 101·83
lambs per 100 ewes.

The writer in *The Times* remarks that
the number of ewes kept for breeding, in all
cases, exceeds the number bred from, which, as
a fact, is doubtless true ; but it would appear
that he is of opinion the proportion of lambs
shown here is, on that account, *lower* than it
should be. Correspondence with breeders, however, has shown me very clearly that, with the
exception of Dorset horn sheep, the number of
ewes in the breeding flock on the 1st of June is
from 7 to 14 % *less* than it is when the ewes are
put to the ram, that is when all the ewes kept
for breeding are accounted for ; and it seems
clear from the separation of " other sheep " in

the official return, that this proportion, say 10%, has still to be *added* to the "breeding ewes" in the official returns

In that case a total of 110,552,405 ewes are engaged in producing these 102,346,808 lambs, and the proportion is 92·58 lambs per 100 ewes, instead of 102 which is the round figure given in *The Times*.

Of course, as the writer of the article rightly says, the actual total of lambs born must be greater than those shown in the tables given, because some die soon after birth and others are sold to the butcher. But in an article published in the *Journal of the Royal Agricultural Society*, 1899, it is shown, by returns supplied by breeders themselves, that for 397 flocks comprising 122,673 ewes, on an average 100 ewes produce 121·48 lambs. (The proportion varies in different pure breeds from 111·1 to 141·77, while for cross-breds the figure given is 129·47.) At this rate the 110,552,405 ewes should produce 134,299,061 lambs, that is 31,952,253 more than the figure given, or 23·79 % of the lambs we have reason to estimate are born.

If we eat anything like 20 % of the animals which do not only provide the breeding stock of the future, but which should, in the future, supply us with mutton, it would appear that breeding for mutton does not pay; or if we eat half these lambs then the death rate must cause a very much more serious loss (12 %) than is generally supposed, and is a matter regarding which breeders would surely welcome investigation by an expert.

From the results of independent observation I am very strongly of opinion that the death rate of lambs is much too high, and that there is reason to think it is increasing, in some breeds and perhaps in some districts markedly so. If our records were reliable such a matter would be made plain, and steps could at once be taken to investigate it, misleading statistics on the other hand are seriously harmful.

Examples can be given for a great variety of

similar facts concerning all kinds of stocks in this country.

Take now another point; the movements of stock are far more important than is generally recognised, and such movements are continually taking place on a large scale. From time to time certain counties or whole districts will show a substantial decrease in the numbers of one kind and increase in another kind of stock. There are reasons for this change which should be understood. At present the necessity governing these movements is bought, at great expense, by the breeder; the facts, whatever they are, which compel him to make a change are gradually forced upon him by the financial results of many years' work. Records of these movements followed by competent investigation would throw much light on the subject, and if, as seems highly probable in some cases, the initial reason therefor is a change in the condition of the land, there at once is opened a wide field for research which could not fail to confer substantial benefit on the industry.

Again, the Royal Commission on horse-breeding has pointed out that the purchase of the best animals by Foreign Governments has swept the country of a large proportion of the best horses, and caused a drain on our horse-breeding stock which requires, most urgently, strong efforts to overcome. Sir Walter Gilbey has written forcibly in the same strain, and the Hunters' Improvement Society has lately taken the matter up. But surely it would not have been possible for us to reach this pass if clear records of the facts had been available from the first. As it is, do we know that a similar drain is not taking place on other classes of stock in the country?

In time of war such records would be of the greatest value. The supply of horses of the stamp required and the districts in which they are to be found would be known. Country depôts could speedily be established for their purchase, in those places best suited for the purpose. The facility with which 'faked' horses

6—2

were disposed of to the War Office during the last war would be checked, and in various ways great saving of money effected.

But the Department should not be content with records obtained in this country alone; imperial as well as foreign records are also necessary, from due consideration of which important benefits would accrue. Canada and Australasia, the United States, Argentina and Germany profoundly affect breeding in this country. The best English-bred stock is always in demand, and the trade to these and other countries is a source of very considerable wealth to us. Thus, the condition of the breeding industry in such countries is of great importance to us. For instance, an article in *The Times*, Sept. 19, 1904, shows a steady reduction in the total of sheep in Great Britain since 1899, when there were 27,238,754, to 1904, when there were 25,207,174, a difference of 2,031,580. From 1903 to 1904 there was a loss of 432,623, and the reduction appears to be going on, for in *The Times* of Oct. 3, 1904, it is recorded of an estate of 12,000 acres, formerly carrying 36,000 breeding ewes, that shortly there will not be a flock left upon it. Now New Zealand, it is stated, has over-sold her sheep, has encroached upon her breeding stock, in which case there must be a greatly reduced supply of mutton from that country for some time; while Australian stations are now being re-stocked and there is doubtless an exceptional demand for sheep in that part of the world. Does it not appear clear that, if the real facts regarding Australasia had been known, the home flocks would have been increased, not decreased, during the last few years, whereas now it is the Argentine or some other country which will probably reap the benefit?

There is still another reason for the collection of records on a broad scientific basis which, in my opinion, is of the greatest importance. It must be remembered that all breeders are essentially experimenters themselves; a great variety of experiments are continually going on all over the country and they are of great value and importance. We want some records of these. Apart altogether from the methods adopted to bring about the results attained, some of which may legitimately be regarded as trade secrets, the results themselves are of the greatest importance. At present, knowledge of some of those which are successful becomes ultimately the general property of the industry, knowledge of others is at best confined to a few individuals, whilst of others again knowledge is lost altogether.

It is, however, not only the successful results which are of value. The bulk of all experiments on all subjects are either only partly successful, or are discarded as failures. In the absence of records these are wholly lost, and yet collectively they are probably of quite as great value as the successes, either as beacons to warn others from unprofitable expenditure, or as hints, connecting links, which may well serve to guide to a successful issue many halting efforts which would otherwise come to nothing.

I would accentuate the part breeders in this

country take in the work of experiment because it appears to me it is truly of the first importance.

By the regulations in force in Germany (a very clear account of which is given by Mr Sedgwick in the *Journal of the Bath and West of England Society*, 1903–4), individual effort is there controlled by a central authority. The result, however commercially successful it may temporarily be, appears to me to be attained at the expense of the most valuable qualities the individual possesses, his own powers of initiation, and one cannot help thinking it is essentially bad policy; at any rate, owing to national characteristics, it is clearly inapplicable to this country.

Experimental farms have been established in various countries, essentially for agriculture, and it is suggested they should be utilized also for experimental breeding. Feeding experiments for milk or meat, certain special problems and test-breeding experiments, can doubtless be

efficiently conducted there; but I would point
out that experimental breeding with large stock,
on such farms, is not possible on a scale which
can be of intrinsic value to breeders generally,
for the following reasons :—The head of stock
on which experiments can be made on such
farms is but small, and sufficient experience of
individual variability cannot be gained there;
the conditions under which stock are kept there
are not the conditions experienced at the hands
of professional breeders; the area of the farm is
restricted; and it is obvious that the conditions
prevailing over such small areas cannot be compared to the highly variable natural conditions
experienced in different parts of the country,
with different sub-soils, at different altitudes,
and with considerable climatic differences. There
are so many and such various factors which influence breeding in different districts that any
conclusions arrived at from experience gained
on these isolated farms would be misleading to
the practical breeder. One might as well expect

the agriculturist to base his scheme of crops
on the results obtained by growing a handful of
grain in a greenhouse, as expect the practical
breeder to organise his business in accordance
with the results to be obtained on an experimental farm.

Thus, in my opinion, the experimental work
of practical breeders is of the first importance;
but inasmuch as their results are not recorded,
are not available for examination and analysis
by those competent to examine and consider
them, the benefits which should accrue from any
of them are reduced to a minimum, while most
of them are wasted and lost altogether.

Finally, there is the advantage to be gained
from accurate records for the regulation and for
the organization of the treatment of disease.
Praiseworthy attempts are indeed made in this
direction, so far as our home stock is concerned,
as the maps which accompany recent Reports
of the Proceedings under the Diseases of Animals Acts show; but, and I imagine the

veterinary surgeon will be the first to agree with me, access to far more complete and various data is necessary to enable the diseases of animals to be fairly treated or efficiently grappled with, and such records are not at present available.

Questions regarding the propriety of permitting, from time to time, the importation of live-stock, come under this head. Our best stock is gradually becoming more and more closely inbred. There are not wanting instances of breeds in this country which, there is strong reason to think, are degenerating for want of fresh blood.

Many of our best animals, year after year, are exported for breeding purposes, and pure strains of our breeds have long flourished in other countries. The time is not far distant when many of our herds at home, if they are to retain their vigour, must be strengthened by new blood which can only be got from abroad. This is the only remedy, and in some cases, it appears to me, the matter is already urgent.

The authorities, in their dread of introducing disease, a very natural fear, are rigidly opposed to the importation of live stock ; but if they had the information necessary to enable them to take a comprehensive view of the situation at home, it would be clear to them that means *must* be devised for enabling a sufficient supply of breeding stock to be imported, and the problem presents no insuperable difficulties.

Briefly, the possession of records such as I would suggest should be obtained, would enable the country to be mapped out in accordance with whatever subject it is desired to investigate. They would enable broad generalizations to be drawn regarding the effects of sub-soil, climate, altitude, water, herbage, etc. on various breeds ; on their fertility and death-rate, on their power to make meat, or bone, or milk, etc.; on their disabilities and diseases; and in these and many other ways focus the main interests of the industry. Such knowledge would enable one to determine the most suitable districts for definite breeds, the natural conditions which are requisite in order to obtain definite results ; they would enable one to grasp clearly the tendencies of any breed to improve or deteriorate, and would point out the lines on which research would clearly benefit breeders. For administrative purposes the value of such records cannot be denied. Finally, imperial and foreign records would keep the breeder informed of the trend of the industry abroad, and enable him to seize opportunities which now are lost to him.

2. *Research.* It is not many years since breeders, as a rule, believed that what they called "nature" was the best doctor for all the evils stock suffer from. Now it is to be hoped they know differently.

There is no radical difference between the effects of domestication on animals and the effects of civilization on man. In both the work of the digestive and excretory systems is modified, the nervous, muscular, vascular and generative systems are affected. In both artificial foods ; excessive exercise or want of exercise ; the strain of artificial life, on any or all the organs of the body; give rise to specialisation. Specialisation in its turn, whether it be in the direction of supreme mental effort, in the production of the largest amount and best quality of milk, or in any other way, creates conditions and results in modifications which are not natural, in the sense in which that word has been used. Thus, the conditions inseparable from domestication are not natural conditions, and " nature " is unable to cure the evils which result. They must be dealt with by those who have studied the laws which govern such matters. They are in fact scientific problems which require for their full elucidation scientific research.

These scientific problems, with which the breeder *must* deal, may be considered under the following heads :

(1) Pathological affections—disease.

(2) The functional modifications due to specialisation.

(3) The results of the forces of heredity and variation.

The two latter are doubtless very closely connected, I have separated them for convenience of treatment.

Efficient control of all these matters presupposes adequate opportunities of research, and, as I have before stated, must be preceded by knowledge which can only be gained from such records as I have already outlined.

(1) Take first, Disease. Animals suffer from disease in a state of nature, and it is possible that very great devastation may result therefrom; but under ordinary circumstances severe loss must be rare, largely because wild animals are rarely confined in large numbers to small areas, and because weakly individuals, prone to disease, die young. Among domesticated animals, on the other hand, disease is common, and both its range and its severity is increased. As in civilized man so in domesticated animals, there is no sufficiently drastic elimination of weakly

individuals, and the crowding together in comparatively small areas is favourable to the production and the spread of disease. These conditions are inseparable from domestication, and it is absolutely necessary that the maximum power of control over disease should be established. For this purpose the veterinary surgeon alone is employed. The Board of Agriculture has a chief veterinary officer, an assistant veterinary officer, and inspectors at ports and for service all over the country; besides which there are veterinary officers attached to certain agricultural societies, and numbers of private practitioners exist, ready to attend to the ordinary diseases of stock; the pathologist even is waiting to be employed in combating the devastations caused by septic organisms.

That disease is not adequately attended to members of the veterinary profession themselves will doubtless admit; but there is organization for the purpose, earnest endeavour is made, and as knowledge increases so the control of disease will become more efficient.

(2)   Take now Specialisation.—The constant
endeavour to improve stock and the pressure
and strain which is put upon animals in order to
reap the greatest possible advantages from them,
results in specialisation ; to this is due not only
the successes but also many of the difficulties
and disappointments breeders meet with. The
power of variation, which all animals and plants
possess, is probably greatly increased by the
conditions of life pertaining to domestication.
The breeder, as well as the horticulturist, is
quick to take advantage of the opportunities
thus offered, and extreme specialisation is the
consequence.   But besides the variation which
is plainly visible, there is correlated with it
variation which is not visible and which is only
indirectly appreciable.   The changes due to the
latter are as sure as in the former case, but they
are not appreciated in their intermediary stages.
They are functional variations essentially affecting
internal organs, their development may be rapid
or slow, but it is not clearly seen, and the result
inevitably is that, when they reach a point which

is appreciable they are already established and the difficulty of overcoming what is evil of them is vastly increased.

The loss resulting from the effect of such specialisation is as great as, indeed, one is inclined to believe, is far greater than any loss from disease. It is the conditions which result from this cause which I find at the bottom of the main difficulties the breeder has to contend with. They are not due to disease, though he very frequently so considers them, they are the natural results of specialisation. They may be evidenced by the gradual increase or reduction in vitality of a definite organ; by hypertrophy or by degeneration of a definite tissue essential for the maintenance of the beast itself, for the purpose for which it is used, or for the propagation of the species. In a word, they are special physiological problems which require the knowledge of an expert to deal with them, and they are of the very greatest importance.

(3) *Heredity and Variation.* Under this

heading we have subjects closely knit together,
a clear knowledge of which is at the foundation
of successful breeding.   The science of heredity
and variation in recent years has made great
strides.   There have arisen schools of experimenters, whose results are gradually but surely
leading to the establishment of laws which will
exercise profound effects, both upon the methods
adopted and the results to be obtained by
breeders.   It is impossible to exaggerate the
practical importance of this work, or to urge too
strongly the advisability of encouraging and
assisting it by every means the breeder has
at his disposal.   Take a solitary instance as
a demonstration of what I mean.   What Sir
Walter Gilbey in his book on Horse-breeding
calls "misfits," he says, always exceed the successful offspring where the aim of the breeder is
high, and the loss which is occasioned thereby
is very great indeed.   These "misfits" are the
result of forces operating on lines which the
science of heredity and variation is engaged in

reducing to law. The art of mating does not supply what is necessary to reduce such failures to their lowest possible proportion, but knowledge of the laws of heredity and variation will enable that to be done, and only by their help can much of the loss now considered inevitable be avoided.

These two classes of problems which confront the breeder, namely the functional modifications due to specialisation and the results of the forces of heredity and variation, are purely physiological matters and can only be dealt with by experts in that science.

At present, the only official to whom the breeder can go for assistance is the veterinary surgeon; he is the only official at the Board of Agriculture or among the officers of any of the Agricultural Societies to whom questions of breeding can be submitted. But such problems as these are not referable to disease, knowledge of them is not included in the veterinary surgeon's studies. He has enough to do to fit himself to combat disease successfully, he cannot be expected to grapple also with the intricacies of a quite independent science. These are special subjects, and inasmuch as they have reference to breeding and include a wide comprehension of the intricate phenomena attending the function of generation, they may be broadly grouped under the term Physiology of the Generative System.

The opportunities I have had of observing the results of practical breeding lead me to believe, that there is no doubt whatever it is precisely these problems which most urgently require attention, that it is the loss which is incurred in consequence of ignorance of these matters which weights and retards the industry, which reduces the profits, swallows up the bonus and prevents breeding from occupying, as it should occupy, a foremost place on the credit side of the national balance sheet.

Consider the question of fertility and the production of healthy offspring, I mean by this the proportion of young stock born and the proportion of those born which are reared. We may say at once there are no official figures available to enable us to arrive at any certain knowledge of these facts; those in authority are not concerned with such matters, they know nothing of them. But there are certain facts which will assist us in arriving at a rough estimate, and these I will give.

Horses. According to the evidence obtained by the Royal Commission on Horse-breeding, 40 % or more of the mares put to the horse failed to produce offspring each year. Then there is the scourge of epidemic abortion, by which there is a continually increasing loss in breeding studs, to what extent it is impossible to say, but the loss is undoubtedly a serious one. Pink-eye, again, causes most serious losses in certain districts, as I know myself; and the mortality of young foals it is quite impossible to estimate, though every breeder suffers loss, and many of them serious loss from this cause also.

Take the breeding mares in this country as 500,000, and say the loss from sterility, abortion, and mortality of the young is equal altogether to not more than 25 % each year, 125,000 foals are thus lost to their breeders annually. The loss in money thus experienced is not represented by the value of the foals at birth, the mare has been unproductive for 12 months, the "service" has been lost, and it must also be remembered that the heaviest loss falls on the best bred animals. I put the value of this loss therefore at an average of £20 per head, and this amounts to £2,500,000 per annum.

Cows. For cattle the only knowledge we have is that which bears on epidemic abortion, a complaint which any herd appears to be liable to and which many appear to suffer from to an extent varying from 12 % to 50 % and over. In a large herd of several hundred cattle with which I am acquainted, where epidemic abortion is not recognized as existing, the loss from abortion appears to average about 10 %.

This herd is exceptionally well managed and great care is exercised with the animals. As for sterility, there is no evidence, but the loss occasioned by cows failing to hold to the bull is very considerable among certain dairy farmers, to whom a continuous crop of calves all the year round is essential for the conduct of their business.

Taking 15 % as the average loss from sterility, failure to hold to the bull, abortion and mortality of calves, a very low estimate; and taking 4,000,000 as the number of cows and heifers breeding ; 600,000 calves are thus lost to breeders annually. The monetary loss thus experienced I calculate to be not less than £7 per head, and the loss thus experienced is £4,200,000 per annum.

Sheep. According to the article already quoted in the *Journal of the Royal Agricultural Society*, 1899, the proportion of ewes which failed to breed or abort their young is 6·76 per cent. The flocks from which these returns were obtained were obviously, as a rule, closely looked after, and it is highly probable the percentage given here is below the average loss experienced throughout the kingdom. On the other hand, it appears that on an average 30 % of the ewes bear twins. Regarding the mortality among lambs experience is much more variable ; a loss of as much as 14 % of the lambs born is regularly experienced in some cases, and occasionally the figure is much higher, but I will put this loss at 10 %. Then taking 11,500,000 as the total for breeding ewes, if all these ewes were fertile, and 30 % of them produced twins, the

lambs born would amount to, say,	14,950,000
but 6·75 per cent of the ewes (776,250) failing to produce young, of which 30 % (232,875) should have borne twins, shows a loss of lambs of	1,009,125
leaving a total of lambs born .	13,940,875
If 10 % of these die . . .	1,394,087
the lambs remaining amount to	12,546,788
and the total loss of lambs per ann. reaches	2,403,212

a loss which at the end of the year will amount to not less than £2,500,000.

The total of such losses for these three classes of stock, therefore, amounts to £9,200,000 per annum.

Impossible as it is to arrive at a true conception, these figures will serve to give a broad idea of the losses breeders experience from these causes alone each year.

I do not of course suggest that any amount of investigation and research will result in wiping out this loss, but I hold a very strong opinion that a few years' work on this special branch of generative physiology, with the facilities a competent officer of a Government Department should possess, would result in knowledge which will enable it to be reduced by half and probably by more than that proportion. For such purpose a competent expert must be employed, the problems requiring solution are essentially physiological problems; it is the functional condition of the generative organs which require investigation; knowledge of the production of the generative products, of their growth and maturation, of the intimate physical relation of the embryo to the mother, and all the intricate physiological phenomena connected with the pre-natal and post-natal nutrition of the embryo that are here concerned; and it is work from which great benefit would undoubtedly be derived.

It may be said, so far as epidemic abortion is concerned, that it lies within the sphere of the veterinary science of the present day; let those who hold this view read the evidence given by veterinary surgeons at the enquiry held by the Royal Agricultural Society in 1894; and if that fails to convince them, let them turn to the Fly-leaf No. 108, issued by the Board of Agriculture in 1904. These matters I propose to deal with in another chapter, here I would only say that this scourge requires attack from various points of view, and that generative physiology is the foundation upon which all such work must be built up.

But such problems as these will be thought

insignificant in comparison with what some may consider to be the main interests of the breeder. For instance, he wants grazing cattle with a frame-work which will enable him to build up a big carcase and quality which will ensure satisfactory meat, without wasting food or time. In spite of the enormous quantities of meat imported into this country there would seem to be always a demand for the best English meat here, but it must be the *best*, and in order to meet that demand it is necessary, not only to feed animals well, but to produce a class of beast which it will pay to feed well. Can anyone say that adequate attention is paid to this point, that there is not great room for improvement in the majority of our beasts; and yet it is this which lies at the root of the grazier's success, it is a breeding problem of the first importance to him.

Again, the far-seeing dairyman wants cows which will yield the greatest amount of milk of the best quality, in winter as well as in summer;

and he wants sure breeding returns.   Until recently, the aim of all cattle breeding in this country was the production of beef.   The dairy cow is a modern development with us, in comparison with those who originally built up the Guernsey and Jersey breeds of cattle.   With comparatively few exceptions the most elementary needs in breeding for the dairy are still not understood.   It is beef which is in the breeder's eye, and he cannot eradicate it.   The result has been a "general purposes cow," a cow which is capable of being fattened for the butcher when it ceases to be used for milk.   It is not physiologically reasonable to suppose you can get, economically, the best milk and a maximum quantity of it, from an animal which is essentially fitted to fatten ; nor can you expect to be able to produce the best meat on a cow, or on the offspring of a cow, specially designed for milking. The "general purposes cow" is really fitted for neither one thing nor the other, and where the *best* meat and the *best* milk is required, as it surely will be if it is not required now, this mongrel is doomed. It seems odd that she should continue to exist; she is kept for milking in a dairy for a number of years, she is expensive to feed as a milker and she gives comparatively poor milk returns; but, apparently because she is worth a few pounds more to the butcher when she is done with for the dairy, she is preferred to a first-rate dairy cow from which much better annual returns can be got; and moreover, she leaves offspring which are, not good.

Similar problems confront the flock-master as regards carcase and quality of meat, but besides these, wool concerns him; and, a very important item, the crop of twins he gets, is of vital importance to him and is a problem of generative physiology which would well repay investigation.

Horse, pig, dog and poultry breeders are confronted with difficulties of the same kind.

Take another branch of enquiry. The effect upon the offspring of various kinds and qualities of food supplied to the pregnant mother. This is a subject of the very first importance, it is difficult to fix a limit to the effect such influence may have. A hint is given of the far-reaching effect of food supplied to the mother by the losses occasioned by "stained" or by "teart" land; the necessity for changing, from time to time, the species of stock grazing on a definite area of land, and in various other ways. But these are only hints; only sufficient is understood to show how important the subject is, to indicate the possibility of insuring the production of better stock, and to emphasise the probability that the good qualities of a strain may be enhanced by the judicious treatment of the pregnant mother, or may be injured or lost by bad treatment.

Great attention has been paid to the feeding of animals for the butcher and for milk, for speed or endurance, but no attention whatever has been paid to the feeding of the pregnant mother of these same animals; and yet the health and vigour of the calf or foal, when born, must have a very great effect on its power for future growth and development, and must very largely depend upon the nutriment supplied to it by the mother.

This question involves a study of nutrition which has hitherto been neglected, of the nutritive value of different grazing lands under a variety of conditions, and of other foods supplied to the mother and by her to her offspring. Why, it may be asked, should land of a special kind produce food particularly favourable to the growth and development of certain breeds. If that were known, there is no reason why the favourable qualities of that land should not be increased and the valuable qualities of that particular breed still further developed. The suggestion opens up a wide field for research, and it is impossible to doubt the primary importance of such work, impossible to deny the influence it must have on the methods and on the success of the breeder.

I have given here but isolated examples of many problems, the solution of which is essential to the satisfactory progress of the industry, and I have referred to them only briefly. It is not possible within the space at my disposal to do more. I will merely add that all such problems are referable ultimately to one or another branch of generative physiology; it must be clear that only experts are capable of dealing with them satisfactorily, and they only when supplied with the knowledge derived from records collected, collated, and analysed on a comprehensive scientific plan.

3. *Administration.* For the full development of the plan here suggested, the Department of Animal Industry would form a section of a new Board of Agriculture, and would be organised and administered by experts with the aid of a practical man of business, familiar with the industry.

It is clear the ultimate details of administration must be determined by the results arrived at by means of the Records and Research work.

The primary administrative work would consist of the organisation of these branches of the Department; subsequently their direction, the publication of their results, the consideration of the numerous representations made to the Department by breeders themselves—intimate relations with whom should be cordially encouraged—the initiation of legislative measures and the ordering of financial matters, would occupy the small staff of competent officials necessary to order and control the whole Department.

The officers of the Administrative Branch would consist of,

(1) A biologist qualified to deal with the Records Department,

(2) A chief veterinary officer,

(3) A man of business, and

(4) The chief of the Department, who would be qualified to organise and direct scientific

research, control the work of the Department
and be directly responsible therefor to the Board
of Agriculture and Fisheries.

These officers would have under their control a sufficient staff to conduct the work of
the Department. This staff would consist of
Collectors of Records, Veterinary Officers and
Scientific Investigators. Of the latter, two or
three specialists would be permanently employed
while others would be temporarily engaged, as
required, for the investigation of special problems.

Thus an intelligence branch would supply
information to the chief of the Department;
after analysis, he would place in the hands of
experts, problems requiring solution ; and the
results of that work would be translated, by
the help of the business official, into methods
capable of practical application.

If the other Departments of the Board were
similarly organized, the President of the Board
would have at his disposal knowledge and advice

which could not fail to transform the Board into a most valuable Government Department.

It will be noted that in the scheme I submit, scientific qualifications are essential for the majority of those who conduct the work of the Department. I make no excuse for that, the Department is concerned with problems which are essentially scientific problems, which require experts to understand and deal with them and, that being so, they are the men who should be employed. As at present constituted, the Board has but the most meagre qualifications for such work.

The late Professor Huxley suggested that modern industry is indeed War in the form of Peace, and needs as surely as war the use of scientific weapons. Such weapons in the case of the breeding industry are trained men of science, and I urge, with the most emphatic earnestness, that the duties the State undertakes with regard to this industry should be entrusted to such men.

CHAPTER IV.

THE BOARD OF AGRICULTURE AND THE BREEDING INDUSTRY.

THE immense value of the breeding industry to the country has been shown, now let us enquire what is the position of the Government as regards that industry.

With perhaps the exception of the Royal Commission on Horse-breeding the Board of Agriculture is the only permanent medium through which its direct efforts are exerted. On the Board there is no one who can claim to be a competent expert on breeding. The subject is practically ignored, attention to such matters is apparently considered outside the scope of the duties of the Board. The actual work done by the Board, so far as animals are

concerned, is practically confined to the prevention of disease. For this purpose the Board
is supplied with a Chief and Assistant Veterinary
Officer, and Inspectors at ports and for service
all over the country. Provision is made to
avoid the introduction of disease by imported
animals, and means taken to stamp out disease
as it occurs at home. Cattle disease, pleuro-
pneumonia, foot and mouth disease and rabies
have been thus successfully eradicated, but
whether the severe loss incurred in bringing
about this result was necessary or not, does not
appear by any means clear.

In the eye of the chief veterinary officer of
the Board (Report of Proceedings under the
Diseases of Animals Act, 1903), it would seem
that the prevention of the introduction or spread
of most diseases, by stopping the importation of
foreign animals and by the slaughter of those of
our own beasts which contract those diseases, are
the chief methods of attacking problems which,
an ordinary person might perhaps be excused

for supposing, members of his profession are ostensibly employed to cure.

On the other hand, sheep scab, glanders, anthrax and swine fever, to which may be added contagious abortion, are still rife; and it would appear, from the remarks of the Assistant Secretary in the Report, that the powers required by the central authorities in order to deal with these diseases, is not considered sufficient. One would like to know how any increased power would be exercised.

The general dissatisfaction with the management and work of the Board of Agriculture is widespread. All over the country one hears from breeders forcible expressions of the uselessness of much of the work done, of the waste of money expended in doing it, and of the impossibility of obtaining from the Board assistance to combat evils which have a really important bearing on the industry.

The late President of the Board apparently recognised that discontent exists, though I doubt if he was at all aware of the extent of it; his method of dealing with the fact, moreover, is hardly calculated to disarm or even to allay criticism. In a speech delivered at a meeting of the Agricultural Organization Society (*Times*, June 21, 1904) Lord Onslow rebukes these discontented ones, and is reported to have said, in reply to their complaints, that "no other industry had at its disposal an expensive organisation maintained at the cost of the State."

It would appear the President held that the mere existence of an expensive organisation is sufficient answer to those who impugn the conduct of its affairs. But critics must surely be acquitted of any charge of cavilling if they insist that this is no answer at all; that the existence of an expensive organisation should, at least, be a guarantee that the claims of those for whom it was created should be adequately attended to; that in point of fact, the President condemned his own Department by his own words.

The breeding industry is increasing in this country. The extent of land now devoted to the growth of cereal crops is upwards of 2,000,000 of acres less than it was 30 years ago, and the number of cattle has enormously increased. Competition has increased also, and the necessity for a wise and broad-minded system, not of control, but of help for this struggling and important industry should be established.

Everyone knows how requests for assistance in large matters are met by the officers of the Department. Money is not forthcoming; the Treasury refuse to supply it; the Board may recognise the importance of the enquiry asked for, or of work required, but unfortunately they have no funds available for the purpose. This expensive organisation is not apparently expensive enough.

Take a single instance—contagious abortion in cattle was recognised as long ago as 1786 as a disease, and is referred to even then as one of the curses of the breeder, but of late years the losses occasioned thereby have been more and more clearly demonstrated by various enquiries made in this and other countries.

In April, 1893, and again in November of that year, the Council of the Royal Agricultural Society called the attention of the Board of Agriculture to the serious losses caused by this disease, and urged the Board, without delay, to make adequate enquiry into the nature and causes of the disease.

In January, 1904, the Council of the Society were informed that, in conjunction with the Lords Commissioners of H.M. Treasury, the Board of Agriculture had considered the matter ; that they did not question the importance of the subject, nor the interest with which it is regarded by agriculturists generally, but that, owing to the need to restrict all expenditure not immediately necessary for the efficient conduct of the public service, their lordships were unable to sanction the necessary financial provision for the enquiry.

In June, 1904, an influential deputation urged
the late President of the Board to institute an
enquiry on the subject. That the matter "was
becoming" a very serious one was not disputed
by him, that the enquiry would take some years,
and that it should probably be conducted by
a qualified specialist, was recognised ; but that,
as several Departmental Committees of the
Board were at present sitting, while it was hoped
that when the labours of one or other of them
were concluded the Board might be able to turn
its attention to this matter, at present the
request was refused.

Now abortion is apparently caused by a
variety of circumstances, few of which are
understood, all of which require investigation.

According to the preliminary returns published
by the Board of Agriculture in 1904 (*Times*,
Aug. 20, 1904) the average number of cattle in
Great Britain for these ten years, 1894–1903,
amounted to 6,594,327. It is now over
11,000,000. Take, as the figures of the Board of

Agriculture quoted elsewhere tend to show, about one-third of these as breeders, say 2,198,109, and that, on an average, not more than 10 % of these abort,—there is good reason to believe the percentage is very much higher—say, 219,810. Taking £5 as the loss occasioned by the abortion of these cows (the lowest estimate given in the Board of Agriculture Leaflet, no. 108), then the loss to the country in these last ten years by abortion of cattle alone, amounts to £10,990,500, and to this must be added the value of the calf the cow should have borne which, at £3 per head, brings the total loss to £17,584,800.

The figures already given for cattle, incomplete as they undoubtedly are, will serve to show how important this question really is to the country; that my estimate is sufficiently conservative I think no one with any knowledge of the ravages caused by the various forms of abortion will deny. But this is not all, mares also suffer and contagious abortion is gaining ground amongst them also. Although of the average number of horses in the kingdom during the last ten years, probably a smaller proportion are breeding animals, the value of their offspring is greater, and the individual loss from abortion of all sorts is a not less serious matter. I do not think we shall err on the extravagant side if we calculate such loss in cattle and horses at not less than £2,500,000 a year, or £25,000,000 sterling during the past ten years.

I leave out sheep, though it is claimed that contagious abortion causes great loss amongst them also in one important breeding district.

It must seem absurd to many to learn that much of this loss has been endured for the want of a few thousand pounds, say £2000 to £5000 a year for five years. Such a sum, spent upon an investigation conducted by qualified experts, would before now have resulted in a wide knowledge of the subject, and a clear idea of the methods to be adopted to eradicate the most serious forms of the disease.

Instead of that, what happens? The Board

of Agriculture issue a leaflet, no. 108, 1904, on Contagious Abortion, which the late President of the Board has referred to as describing a "remedy" for the disease. It will be instructive to examine what is there set down, the value of the advice there given, and whether or not the President of the Board was justified in using the term "remedy" in connection with that advice.

In the leaflet it is stated, the results embodied therein are the results of experimental enquiries carried out by Professor Cave at the request of the Board.

Abortion is divided into :

1. Simple or Sporadic,
2. Specific or Contagious.

The former, very incompletely and inadequately described, is dismissed in a few words. Contagious abortion is then defined, and Professor Cave's experiments and treatment for "stamping out the disease" given in detail.

These experiments consist of A. the internal administration of carbolic acid ; B. the external use of antiseptics; and C. the combined treatment of A. and B. The conclusions he arrived at were, that the internal administration of carbolic acid fails to prevent contagious abortion, but that the external use of antiseptics was entirely successful. The advice given in the leaflet, then, resolves itself into advice concerning the prevention of contagious abortion in cattle by means of disinfection, and in a final section a summary of the treatment advised is given.

Information I have received from breeders shows that from 10% to 12% of abortion is not unusual in a herd in which no contagious abortion is proved to exist, and it is important to bear this fact in mind, in considering Professor Cave's results, for he offers no certain evidence that the herds he treated actually suffered from contagious abortion.

The figures given would appear clearly to demonstrate that great value is to be derived from his external antiseptic treatment, and, if the animals he treated *were* suffering from contagious abortion, there would, from his statements, appear to be no doubt of the benefits his treatment conferred.

But there is another point. In paragraph 2 of the Summary of Treatment it is advised that "no aborted cows should be bred from, but should be fattened and sold"; while further, an apparently generous concession is made for the herds experimented upon,—it is repeated for each experiment,—viz. that "no restrictions were placed on the sale or purchase of cows for the maintenance of the herd."

One can hardly be blamed if these statements suggest a possibility that cows which had aborted were removed from the herds experimented upon. It is almost inconceivable, however, that this course should have been adopted in these herds, without due note having been made of the fact, for if so, it is clear that the chief part of the value claimed for the use of antiseptics, as described in this leaflet, is unjustifiably claimed.

But assuming that the cattle treated *were* suffering from contagious abortion, and assuming that aborted cows *were* retained in the herds experimented upon, the benefit derived from the leaflet resolves itself into, not a cure for the disease, but a method for preventing the spread of it.

The problems regarding this disease which breeders require solved, are :

1.   The origin of the disease,

2.   The means by which inoculation is brought about,

3.   The means by which the disease can be diagnosed in time to effect,

4.   The cure of it.

This leaflet states that, in 1893 the disease was first brought to the attention of the Board of Agriculture,—if they had had a reasonably competent staff they would have known of it very many years before that date ;—in it, there is no evidence advanced to demonstrate the origin of the disease, none to show how it is primarily contracted, no method of diagnosing it in a pregnant animal is referred to, and no method of curing it established. Instead, after seven years' work, it has been attempted to show how the spread of a contagious disease, the origin and progress of which is not understood, is to be prevented ; and it is finally declared that an animal which once contracts the disease is unfit for breeding.

This is surely a travesty of scientific method. It cannot indeed be wondered at if breeders are dissatisfied with such results and such advice, emanating from their Board of Agriculture; and it must be admitted, that the reference to this Leaflet by the late President of the Board, as containing "suggestions for a remedy," is not calculated to engender confidence in the ability of the Board, as at present constituted, to grapple with important problems which require for their solution scientific treatment and research.

That such want of confidence is widespread there is not wanting ample evidence. A recent memorial to the lords of the Treasury regarding another section of the duties assigned to the Board, is proof of that; and yet it is precisely scientific research which is required for the elucidation of almost all the problems with which the Board of Agriculture and Fisheries should concern themselves.

At the Anniversary Meeting of the Royal Society, Nov. 30, 1904 (*Nature*, Dec. 1, 1904), the President of the Society remarked, in connection with the welfare of the country, "the expenditure of the Government on scientific research and scientific institutions, on which its commercial and industrial prosperity so largely depend, is wholly inadequate in view of the present state of international competition."

Such a statement is certainly applicable to the Board of Agriculture. It may be claimed it is not constituted for such work, that its organisation does not admit of it; in that case we may ask why that is so, since breeding plays such an important part in the economy of Agriculture, and since science is essential for its progress.

It will not perhaps be inappropriate to glance for a moment at the cost to the country of this Department, and incidentally to observe the principles which guide the disposal of the money granted for the prosecution of its work.

According to the statement of the Civil Service and Revenue Department, Appropriation Accounts for 1902–1903, Vote XI.

	£	s.	d.
A total sum was granted to the Board of . . .	120,119	0	0
to which must be added balance on hand from 1901–2 Diseases of Animals Fund £4299. 15s. 6d.			
Salvage . £342. 6s. 1d.	4,642	1	7
making a total of . . . £124,761 1 7			

From this must be deducted
Balance in hand 1902–3
For the Board,

Vote XI. £5,451. 11s. 11d.

Diseases of Animals			
Fund . £1,617. 14s. 2d.	7,069	6	1

leaving a total to be accounted
for of £117,691 15 6

It is accounted for thus :

The Diseases of Animals Expenses.

	£	s.	d.	
Salaries and fees	18,581	16	11—	40·51 %
Travelling expenses	9,510	2	7—	20·73 %
Compensation for slaughtered animals	11,340	12	8—	24·72 %
Miscellaneous expenses	6,441	15	3—	14·04 %
Total	£45,874	7	5—	100 %

That is 38·98 % of the whole grant.

The remainder of the grant was expended thus:

	£	s.	d.	
Salaries and wages	48,657	5	3—	67·75 %
Travelling expenses	4,652	13	10—	6·48 %
Statistics . . .	4,559	1	9—	6·35 %
Education . .	8,900	0	0—	12·39 %
Miscellaneous expenses	5,048	7	3—	7·03 %
Total	£71,817	8	1—	100 %

That is 61·02 % of the whole grant.

Total accounted for £117,691. 15s. 6d.

Compare this statement with the figures

given in the United States Year Book of the
Department of Agriculture, 1902. Omitting the
Agricultural Experimental Stations Grant, which
is a separate grant, and the office expenses
thereof, and the Weather Bureau.

Total allowed for year 1903	$3,674,200

The Department of Animal Industry
  receives of this sum .   .   .   1,660,000

That is 45·18 % of the whole grant.

    The remainder is appropriated as follows:

Salaries	$465,500—	23·11 %
Contingent expenses .	43,000—	2·14 %
Collection of Agricultural statistics . . .	94,200—	4·68 %
Library and publications	212,000—	10·52 %
Purchase and distribution of seeds . . .	270,000—	13·40 %
Scientific investigations and experiments on a great variety of subjects .	929,500—	46·15 %
Total	$2,014,200—	100 %

That is 54·82 % of the whole grant.

Total accounted for $3,674,200.

I find no record of the details of the expenditure of the Department of Animal Industry, but the volumes published by that Department are ample evidence of the work done. Examination of these records will show that salaries, travelling expenses, compensation, and miscellaneous expenses do not, as is the case with our Board, comprise the sole heads of expenditure here. Indeed, to whatever extent this Department is capable of improvement, it is impossible to deny that, apart altogether from the amount spent, the attention our Board of Agriculture pays to animal industry is in no way comparable to the broad views which govern the principles of the Americans.

For the rest, that is, the General Expenditure of the Department, the contrast is sufficiently startling.

Where our Board spends in salaries 67·75 % America spends 23·11 %. Our "travelling expenses" is not represented in the American balance sheet. Our miscellaneous expenses, 7·03 %, compare with America's contingent expenses $2\cdot14\%$; where our statistics cost $6\cdot35\%$ America's cost $4\cdot68\%$; finally, while the $12\cdot39\%$ we spend on education may compare with the American item, library and publications, $10\cdot52\%$, there remains for the Americans $13\cdot40\%$ for the distribution of seeds, and the solid item, scientific work, $46\cdot15\%$.

Regarding the actual sums expended in each country under these different heads, it is clear no comparison can be made. The same objection, however, will not so completely hold for a comparison of proportionate expenditure, though it is also clear, any comparison of fixed charges, where the total sums involved are so widely different, cannot be fairly stated in percentages.

Broadly, however, such comparison shows that, whereas, of the whole grant, our Board expends on the Diseases of Animals $38\cdot98\%$ the Americans spend on the Department of Animal Industry a similar proportion, though for very different work . . . . $45\cdot18\%$

Assuming that the expenditure of our Board under this head is as satisfactory as an analysis of the expenses of the Department of Animal Industry would show, though with the volume before us it is impossible this should be so ; when the remainder of the grant is considered, wide divergence is observed.

The figures I have given show this sufficiently plainly, but I will draw particular attention to the most conspicuous item.

It appears that while our Board spends on salaries .   .   .   .   67·75%
the Americans spend   .   .   .   23·11%
and consequently it is clear that if we spent in salaries, for this portion of the Board's work, the same proportion as the Americans thus spend, we should have for disposal   .   .   .   .   44·64%
of this portion of the grant; which represents a sum of £32,059.

While making substantial allowance for the increased proportionate cost of administering our small grant, in comparison with that incurred by the Americans, the very wide difference here shown strongly suggests that the item for salaries is quite out of proportion to the work done. It must be remembered that the numerous scientific investigations carried out by the American department require many expert officers who must be paid.

The American Grant is distributed on quite different lines to our own, and if it is very much larger, very many more officers are required to do the work. The allowance to be made for the increased proportionate cost of administration is, therefore, not so great as might at the first glance be supposed.

It is impossible to deny that this country requires, for the satisfactory progress of every branch of the agricultural industry, scientific investigation and experiment, quite as much as do the Americans; indisputably, it is precisely such work which is really urgently needed in England. With one half of £30,000 a year a large amount of most valuable work could be done. It would, I believe, when supplemented with the amount already expended on the Diseases of Animals, go far to finance such a Department of Animal Industry as I have already outlined. Is it not evident that it is not, after all, so entirely the want of money which is the stumbling-block, as the manner of its appropriation?

We have, in fact, a Board of Agriculture which is not competent to undertake the work required by those of our agriculturists who are breeders. How absurd is such an arrangement.

A matter; sheep-scab, foot-rot, abortion*, etc., arises, requiring immediate expert advice. After numerous memorials and years of intolerable

* Since the above was written I see it is recorded at a meeting of the Central Chamber of Agriculture, that the new President of the Board of Agriculture has now appointed a Departmental Committee on Abortion; it appears however that the Chamber is not satisfied with the duties assigned to that Committee. (*Times*, May 3, 1905.)

The fact that the Board has at last yielded to the pressure brought to bear upon it, does not in any way affect what I have written.

delay, perhaps a Royal Commission or Departmental Committee is appointed to enquire into the evil, which in the meantime has resulted in irrevocable and severe loss. When the cost of the Commission, the cost of the steps they advise to be taken, and the cost of the delay is added up, is it not clear that the cost to the country, entailed by this procedure, must be infinitely greater than would be the annual cost of a Department which is itself constituted so as to enable it to grapple with such problems at their outset.

One hesitates to describe in fitting terms the condition thus disclosed, but one may perhaps be permitted to agree with the late President of the Board, that his department is indeed a very "expensive Department" to the country.

A word may be said regarding the grants made for Agricultural Education and Research, by the Board of Agriculture. The Report (1903) shows how little attention is paid to stock. In one farm-school (Cumberland and Westmoreland) 3 hours a week are devoted to farm anatomy and physiology; while dairy and poultry instruction, and here and there practical experience in the handling and the care of stock, satisfies the rest. Experiments on feeding are conducted at the University College of North Wales and the Durham College of Science; on sheepbreeding, at the University College of North Wales; on the composition and production of milk, at Cambridge University and Reading College; whilst grants of £200 for experimental research on the origin and cause of flavour in dairy produce, and £225 for experimental research on manure and mutton, a total of £425, distributed between three Agricultural Societies and one University, is apparently all the money available for experimental work.

Something of course is done by the public themselves for the breeding industry; notably the Societies for the improvement of different breeds do good work, and efforts have been and are being made by Agricultural Societies to further the interests of certain classes of breeders, or of breeders in certain districts.

Research is occasionally indulged in. The Royal Agricultural Society some years ago issued an important paper on the Lambing Pen; the Bath and West of England Society carried on for some years an enquiry regarding "teart" land, which gravely affects breeding stock ; the Royal Agricultural Society interested itself some time ago in an enquiry regarding the fertility of Sheep, and the Highland and Agricultural Society is now itself officially engaged in a similar enquiry.

Apart, however, from problems directly concerned with the products derived from stock, there is little organised work done, and to problems which directly concern breeding, scarcely any attention is paid. Doubtless there are other examples of work similar to those I have given, but they are all isolated, spasmodic efforts, and for that reason are of limited value.

In such matters there has been no broad attempt at coordination, no systematic effort to obtain a clear and comprehensive knowledge of the conditions, throughout the kingdom, affecting any of the great problems which the breeder has to face : and yet it is only by such methods that really valuable work can be done.

A scheme is now on foot, originated by the Hunters' Improvement Society, to initiate a "serious and sustained effort...to organize the industry, to help the breeder to increase and encourage the production of hunter-bred sires, to devise methods of rendering assistance in breaking and selling horses, and to promote co-operation among the producers themselves." It is suggested that the Hunters' Improvement Society, the Royal Commission on Horse-Breeding, the Brood Mare Society, the Board of Agriculture, the War Office, and the Department of Agriculture and Technical Instruction for Ireland, should combine to form a permanent Committee for this purpose.

A report of a committee of the Hunters'
Improvement Society, on the breeding of hunters
and general utility horses, which resulted in the
above scheme, draws attention to the "almost
entire want of system and scientific treatment
in the production of light horses of the hunter
type." They point out, as has long been known,
that the country has been and is being drained,
to a large extent, of its best mares by purchases
from abroad, and that there is great need for
some organized effort to retain in the country
the good brood mares that are left. (*Times*,
Dec. 9, 1904.)

It is to be hoped this serious effort will be
sustained, and that it will prosper, but even so it
is only breeders of this class of horse who will
benefit at all ; and if the breeders of every
other class followed suit, such work as is
indicated above would barely touch the fringe of
the great issues which bind down the industry
in its present unsatisfactory state.

For some time periodicals dealing especially
with agriculture, and the press generally, have

published papers and letters on organisation. The Agricultural Organisation Society has been founded. Organisation has been held up to us as the panacea for all agricultural ills, it has become a catch-word, the essential meaning of which has been lost or forgotten. Undoubtedly, proper organisation is at the root of the matter, nothing great can be done without it ; whether we attempt to deal with agriculture proper or with breeding, whether we concern ourselves with grain or meat, or any other problem, on a sufficiently broad scientific basis, organisation is essential. But organisation, that is the connection and co-ordination of parts for vital functions, requires a centre with sensory and motor attributes, a head which is capable of receiving impressions and capable of directing energy, on lines which shall be advantageous to the whole organised body. Without this head, organisation is a farce, and in the same way, the exertions of many organisations in the concern of one great industry results in a loss, not a gain of strength, in friction, confusion and distress.

In dealing with agriculture and breeding we are dealing with what is, for many reasons, one of the main interests, not only of the British Islands but of the Empire. The head of an organisation which deals adequately with these subjects, must have many qualifications. It must have the power to enable it to gain a thorough knowledge of the condition of the industry throughout our own kingdom, in our colonies and dependencies, and in foreign countries. It must have, moreover, the wit to understand the bearing of the information so gained, the expert knowledge which is essential to that understanding, and the power of authoritatively presenting such information to those concerned. Thus, it must be willing to listen to the troubles and difficulties agriculturists and breeders meet with in this country, to be able intelligently to advise, to promote legislation, and, as required, to initiate and carry out research ; for all of which purposes the expert is essential.

These, briefly and in the broadest sense, are

the essential preliminary requirements for agricultural and breeding organisation. The interests are so vast and so diverse, it is impossible that complete machinery for such purpose should be supplied by any combination of private individuals or of societies, impossible that the whole of the necessary power should be conferred on any body so organised; it is the concern of the State.

This fact was apparently recognised in 1889 when the Board of Agriculture was established, but however fitted it was in those days to carry out the work for which it was designed, it is certainly incompetent to supply present-day needs.

Such efforts as the Board has made to keep pace with the modern needs of breeders are wholly inadequate. The result is that speeches, made at the council meeting of an influential society, deriding the capabilities of a government department to deal with a crying evil which is ostensibly within the sphere of its legitimate work, are abundantly justified.

To a private individual such a position would be an ignominious one, and any business concern, the affairs of which were conducted in such manner, would certainly fail.

An antiquated government department only avoids liquidation because it has no competitors, but the *failure* is not thus avoided, it falls quite surely and with no less force on the country.

I have endeavoured, in these pages, in spite of considerable difficulties, to put before my readers a few facts regarding the present position of a great industry; I have attempted to indicate the nature of the work legitimately required to foster its growth, and to support breeders in this country in their efforts to retain their supremacy in the world.

If I am right in my conclusions,—it is clear that only a government department can do what is primarily required for the breeding industry ;

it is clear that at present a government department stolidly blocks the plainly defined road of progress; and, if such obstruction is to be removed, it is also clear the Board of Agriculture and Fisheries must be reorganised on a broad scientific basis.

INDEX.

Printed in the United States
By Bookmasters